INFINITE SERIES AND

$$\sum_{n=1}^{\infty} \frac{1}{n^2} = \frac{\pi^2}{6}$$

$$\prod_{n=1}^{\infty} \left(1 + \frac{1}{n^2}\right) = \frac{e^{\pi} - e^{-\pi}}{2\pi}$$

1) A concise and complete introduction to the theory of Infinite Series and Products.

2) An excellent supplementary text for all Mathematics, Physics and Engineering students.

3) 63 illustrative examples and 176 characteristic problems to be solved.

4) Odd numbered problems are provided with answers.

About the Author

Demetrios P. Kanoussis, Ph.D

Kalamos Attikis, Greece

dkanoussis@gmail.com

Dr. Kanoussis is a professional Electrical Engineer and Mathematician. He received his Ph.D degree in Engineering and his Master degree in Mathematics from Tennessee Technological University, U.S.A, and his Bachelor degree in Electrical Engineering from the National Technical University of Athens (N.T.U.A), Greece.

As a professional Electrical Engineer, Dr. Kanoussis has been actively involved in the design and in the implementation of various projects, mainly in the area of the Integrated Control Systems.

Regarding his teaching experience, Dr. Kanoussis has long teaching experience in the field of Applied Mathematics and Electrical Engineering.

His original scientific research and contribution, in Mathematics and Electrical Engineering, is published in various, high impact international journals.

Additionally to his professional activities, teaching and research, Dr. Kanoussis is the author of several textbooks in Electrical Engineering and Applied Mathematics. A list of his publications is shown below:

Mathematics Textbooks

1) Sequences of Real and Complex Numbers, e-book, March 2017.

2) Infinite Series and Products, e-book, April 2017.

3) Complex Numbers, an Approach of Understanding, e-book, June 2017.

4) Polynomial Equations, e-book, May 2017.

5) Challenging Problems in Trigonometry, e-book, March 2015.

6) Topics in Applied Mathematics, paperback, November 2011, (Greek Edition).

Electrical Engineering Textbooks

1) Introduction to Electric Circuits Theory, Vol. 1, e-book, May 2017.

2) Direct Current Circuits Analysis, Vol. 2, e-book, May 2017.

3) Introduction to Electric Circuits Theory, paperback, August 2013, (Greek Edition).

Infinite Series and Products.

 Author: Demetrios P. Kanoussis.

Inquires should be addressed directly to the author,

Demetrios P. Kanoussis

dkanoussis@gmail.com

First edition, April 2017

Preface

This book is a complete and self contained presentation on the fundamentals of **Infinite Series and Products** and has been designed to be **an excellent supplementary textbook** for University and College students in all areas of Mathematics, Physics and Engineering.

Infinite Series and Products constitute an important division of Applied Mathematics with an enormous range of applications in various areas of Applied Sciences and Engineering.

The Theory of Infinite Series and Products relies heavily on the Theory of Infinite Sequences and therefore the reader of this text is urged to refresh his/her background on Sequences and related topics.

In our e-book "**Sequences of Real and Complex Numbers**" the reader will find an excellent introduction to the subject that will help him/her to follow readily the theory developed in the current text.

The content of this book is divided into 11 chapters.

In **Chapter 1** we introduce the Σ and the Π notation which is widely used to denote infinite series and infinite products, respectively.

In **Chapter 2** we present some basic, fundamental concepts and definitions pertaining to infinite series, such as convergent series, divergent series, the infinite geometric series, etc.

In **Chapter 3** we introduce the extremely important concept of **Telescoping Series** and show how this concept is used in order to find the sum of an infinite series in closed form (when possible). In this chapter we also present a list of **Telescoping Trigonometric Series**, which arise often in various applications.

In **Chapter 4** we develop some **general Theorems** on Infinite Series, for example deleting or inserting or grouping terms in a series, the Cauchy's necessary and sufficient condition for convergence, the widely used necessary test for convergence, the Harmonic Series, etc.

In **Chapter 5** we study the **Convergence Test for Series with Positive Terms**, i.e. the Comparison Test, the Limit Comparison Test, the D' Alembert's Test, the Cauchy's n^{th}Root Test, the Raabe's Test, the extremely important Cauchy's Integral Test, the Cauchy's Condensation Test etc, and show how these tests are applied in practice in order to establish the convergence or the divergence of a given series.

In **Chapter 6** we study the **Alternating Series** and the investigation of such series with the aid of the Leibnitz's Theorem.

In **Chapter 7** we introduce and investigate the **Absolutely Convergent Series and the Conditionally Convergent Series**, state some important Theorems on Absolute and Conditional Convergence and define the Cauchy Product of two absolutely convergent series.

In **Chapter 8** we give a brief review of **Complex Numbers and Hyperbolic Functions**, needed for the further development of series theory from real to complex numbers. We define the Complex Numbers and their Algebraic Operations and give the three representations i.e. the Cartesian, the Polar and the Exponential representation of the Complex Numbers. The **famous Euler's Formulas** and the important De Moivre's Theorem are presented and various interesting applications are given. In this chapter we also define the so called Hyperbolic Functions, i.e. the Hyperbolic cosine (cosh x), the Hyperbolic sine (sinh x), the Hyperbolic tangent (tanh x) and the Hyperbolic cotangent (coth x) and state various identities among them. **Finally we extent the definition of the Trigonometric and the Hyperbolic Functions from Real to Complex values of their arguments.**

In **Chapter 9** we introduce the theory **of Series with Complex Terms**, define the convergence in the complex plane and present a few important Theorems which are particularly useful for the investigation of series with complex terms.

In **Chapter 10** we define the **Multiple Series** and show how to treat simple cases of such series.

In **Chapter 11** we present **the fundamentals of the Infinite Products**, give the necessary and sufficient condition for the convergence of Infinite Products, define the Absolute and Conditional Convergence of Products and investigate various types of Infinite Products with the aid of the Theorems developed.
In particular, in this chapter we present the well known "**Euler's product formula**" for the $(\frac{\sin x}{x})$ function and show how Euler used this product to evaluate the infinite series $\frac{1}{1^2}+\frac{1}{2^2}+\frac{1}{3^2}+\cdots$, (**the famous Basel's Problem**).

The **63 illustrative Examples** and the **176 characteristic Problems** to be solved are designed to help students develop a solid background, broaden their knowledge and sharpen their analytical skills on the subject.
A brief **hint** or **detailed outline** of the procedure to be followed, in solving more complicated problems is often given.
Finally answers to odd numbered problems are also given, so that the students can verify the validity of their own solution.

Demetrios P. Kanoussis

Table of contents

1. The Σ and Π notation .... **page 10**

(Definition and various simple Examples involving the $\boldsymbol{\Sigma}$ and the $\boldsymbol{\Pi}$ notation).

2. Basic Concepts and Definitions .... **page 12**

(Convergent and Divergent Series, sequence of partial sums, the infinite geometric series).

3. Telescoping and Related Series .... **page 19**

(Telescoping sums and Telescoping series, case where the general term of the series is $u_n = f(n+1) - f(n)$, or case where $u_n = af(n) + bf(n+1) + cf(n+2)$ with $(a+b+c) = 0$, Trigonometric Telescoping series such as $\sum_{n=1}^{\infty} Arccot(n^2+n+1)$ or $\sum_{n=1}^{\infty} Arccot(2n^2)$, etc).

4. General Theorems on Series .... **page 31**

(Deleting, inserting or grouping terms in a convergent series, the Cauchy's necessary and sufficient condition for convergence, the necessary condition for convergence, the divergence of the Harmonic series).

5. Convergence Criteria for Series with Positive Terms... **page 41**

(The Comparison Test, the Limit Comparison Test, the D' Alembert's Test, the Cauchy's n^{th} root Test, the Raabe' Test, the Raabe's generalized Test, the Cauchy's Integral Test, the Cauchy's Condensation Test, the $p-$ series).

6. Alternating Series .... **page 71**

(The Leibnitz's Test for Alternating Series, the Alternating Harmonic Series, the Alternating $p-$ Series).

7. Absolute and Conditional Convergence .... **page 76**

(Absolutely and Conditionally convergent series, important Theorems and applications, rearrangement of Absolutely Convergent Series, the Cauchy product of Absolutely Convergent Series).

8. Complex Numbers and Hyperbolic Functions ....**page 89**

(Complex Numbers, operations with complex numbers, Cartesian, Polar and Exponential form of complex numbers, the De Moivre's Theorem and applications, Hyperbolic functions of real argument, Trigonometric and Hyperbolic functions of complex argument, the famous Euler's formulas, the exponential function e^z).

9. Series with Complex Terms ....**page 110**

(Convergence of series with complex terms, important Theorems, absolute convergence of complex series and applications, periodicity of the exponential function e^z,summation of Trigonometric series of the form $\sum c_n \cos(n\phi)$ or $\sum c_n \sin(n\phi)$ with the aid of series with complex terms).

10. Multiple Series ....**page 119**

(Multiple sums and multiple series, absolute convergence in multiple series and applications, important Theorems).

11. Infinite Products ....**page 124**

(Convergent and Divergent infinite Products, Necessary condition for convergence, absolute and conditional convergence of infinite products and related important convergence Theorems, the Euler's product formula for the $\left(\frac{\sin x}{x}\right)$ function, the famous Basel's Problem and Euler's brilliant solution).

1. The Σ and Π Notation.

Let $x_1, x_2, x_3, \cdots, x_n$ be n numbers, (real or complex).

The sum $(x_1 + x_2 + x_3 + \cdots + x_n)$, can be expressed briefly using the **Greek symbol Σ** ,as

$$\sum_{k=1}^{n} x_k = x_1 + x_2 + x_3 + \cdots + x_n. \quad (1\text{-}1)$$

The index k is a **dummy variable**, in the sense that

$$\sum_{k=1}^{n} x_k = \sum_{m=1}^{n} x_m = \sum_{i=1}^{n} x_i = x_1 + x_2 + x_3 + \cdots + x_n.$$

The sum in (1-1) could also be written as,

$\sum_{k=2}^{n+1} x_{k-1}$ or as $\sum_{m=3}^{n+2} x_{m-2}$ or as $\sum_{i=4}^{n+3} x_{i-3}$, etc.

Similarly, the product $(x_1 \cdot x_2 \cdot x_3 \cdots x_n)$ can be expressed briefly using the **Greek symbol Π** as,

$$\prod_{k=1}^{n} x_k = x_1 \cdot x_2 \cdot x_3 \cdots x_n. \quad (1\text{-}2)$$

Obviously, k is a **dummy variable**, again, since

$$\prod_{k=1}^{n} x_k = \prod_{m=1}^{n} x_m = \prod_{i=1}^{n} x_i = x_1 \cdot x_2 \cdot x_3 \cdots x_n.$$

The product in (1-2) could also be written as

$$\prod_{k=2}^{n+1} x_{k-1} \; or \prod_{m=3}^{n+2} x_{m-2} \; or \; \prod_{i=4}^{n+3} x_{i-3} = x_1 \cdot x_2 \cdot x_3 \cdots x_n.$$

It is easy to show that, if c and d are any two constants, then

$$\sum_{k=1}^{n} (c \cdot x_k + d \cdot y_k) = c \cdot \sum_{k=1}^{n} x_k + d \cdot \sum_{k=1}^{n} y_k \, . \quad (1\text{-}3)$$

PROBLEMS

1-1) Show that $\sum_{k=1}^{n} k = 1 + 2 + 3 + \cdots + n = \frac{n\cdot(n+1)}{2}$

1-2) Show that $\sum_{k=1}^{n} k^2 = 1^2 + 2^2 + 3^2 + \cdots + n^2 = \frac{n\cdot(n+1)\cdot(2n+1)}{6}$

1-3) Show that $\sum_{m=1}^{n} m^3 = 1^3 + 2^3 + 3^3 + \cdots + n^3 = \left\{\frac{n\cdot(n+1)}{2}\right\}^2$

1-4) If c is any constant, $(c \neq 0)$, show that $\prod_{i=1}^{n}(c \cdot x_i) = c^n \cdot \prod_{i=1}^{n} x_i$.

1-5) Show that $\prod_{n=0}^{k} \cos(2^n x) = \frac{1}{2^{k+1}} \cdot \frac{\sin(2^{k+1}x)}{\sin x}$, $(x \neq 0)$.

Hint: Make use of the Trigonometric identity, $\mathbf{\cos x = \frac{\sin 2x}{2\sin x}}$.

2. Basic Concepts and Definitions.

Let $(u_n) = \{u_1, u_2, u_3, \cdots, u_n, u_{n+1}, \cdots\}$ be a given sequence. From this given sequence (u_n), we may form a new sequence $(s_n) = \{s_1, s_2, s_3, \cdots, s_n, s_{n+1}, \cdots\}$, in the following way:

$$s_1 = u_1$$

$$s_2 = u_1 + u_2 = \sum_{k=1}^{2} u_k$$

$$s_3 = u_1 + u_2 + u_3 = \sum_{k=1}^{3} u_k$$

$$\vdots$$

$$\vdots$$

$$s_n = u_1 + u_2 + u_3 + \cdots + u_n = \sum_{k=1}^{n} u_k$$

The sequence (s_n) is called the sequence of the partial sums of (u_n) .The number s_n is the $n^{\underline{th}}$ partial sum of(s_n).

As n keeps increasing, the number of terms of (u_n) included in the summation, keeps increasing as well, and as $n \to \infty$, all the terms of (u_n) are being summed up, i.e.

$$\lim_{n\to\infty} s_n = \lim_{n\to\infty}\left(\sum_{k=1}^{n} u_k\right) =$$

$$= \sum_{k=1}^{\infty} u_k = u_1 + u_2 + u_3 \cdots + u_k + u_{k+1} + \cdots \qquad \text{(2-1)}$$

The expression $\sum_{k=1}^{\infty} u_k$ is called **an infinite series**, or just for brevity, **a series. The term u_k is the $k^{\underline{th}}$ term, or the general term of the series.** Obviously,

$$\sum_{k=1}^{\infty} u_k = \sum_{n=1}^{\infty} u_n = \sum_{m=1}^{\infty} u_m = u_1 + u_2 + u_3 \cdots + u_k + u_{k+1} + \cdots$$

Definition 2-1: If the sequence (s_n) of partial sums, converges to **a finite limit** S, we say **that the series $\sum_{n=1}^{\infty} u_n$ is convergent, and S is the sum of the series.** In symbols,

$$\lim_{n\to\infty} s_n = S \Leftrightarrow \sum_{n=1}^{\infty} u_n = S. \qquad (2\text{-}2)$$

If the sequence of partial sums (s_n) diverges, the series $\sum_{n=1}^{\infty} u_n$ is called **divergent**.

A divergent series could tend to $+\infty$ or to $-\infty$, or it could have no limit at all.

For brevity and notation economy, we may sometimes write $\sum u_n$ instead of the full notation $\sum_{n=1}^{\infty} u_n$, if this causes no ambiguity.

Given a series, there are two fundamental questions to be answered,

a) Is the given series convergent or divergent, and

b) In case the series is convergent, what is its sum?

Both questions, in general, are difficult to be answered. In practice, answers to the aforementioned questions, are obtained by means of Theorems and Techniques, to be developed in the rest of this book.

For example, as we shall show, the so called **Harmonic Series** $\sum_{n=1}^{\infty} \frac{1}{n}$ diverges to $+\infty$, i.e.

$$\sum_{n=1}^{\infty} \frac{1}{n} = \frac{1}{1} + \frac{1}{2} + \frac{1}{3} + \cdots + \frac{1}{n} + \cdots = +\infty,$$

while the series

$$\sum_{n=1}^{\infty} \frac{1}{n^2} = \frac{1}{1^2} + \frac{1}{2^2} + \frac{1}{3^2} + \cdots + \frac{1}{n^2} + \cdots < \infty$$

The notation $\sum_{n=1}^{\infty} \frac{1}{n^2} < \infty$ means that the series **converges to a finite number**. This last series, is related to the famous **"Basel's Problem"**, solved for the first time in 1753, by **Leonard Euler**, one of the greatest Mathematicians of all times, who proved the astonishing and unexpected result, that

$$\sum_{n=1}^{\infty}\frac{1}{n^2}=\frac{\pi^2}{6}.$$

As another Example the series $\sum_{n=1}^{\infty}\frac{1}{\sqrt{n}}=\frac{1}{\sqrt{1}}+\frac{1}{\sqrt{2}}+\frac{1}{\sqrt{3}}+\cdots=+\infty$ (i.e. diverges to $+\infty$), while the series $\sum_{n=1}^{\infty}\frac{\sin(n+1)}{n^3}<\infty$, (converges to a finite number).

Note: The series $\sum u_n$ is sometimes called **an arithmetic series**, since its terms u_n are just numbers. There are cases where $u_n = u_n(x)$, and in such cases

$$\sum_{k=1}^{\infty} \boldsymbol{u_k}(\boldsymbol{x}) = \boldsymbol{u_1}(\boldsymbol{x}) + \boldsymbol{u_2}(\boldsymbol{x}) + \boldsymbol{u_3}(\boldsymbol{x}) + \cdots + \boldsymbol{u_n}(\boldsymbol{x}) + \boldsymbol{u_{n+1}}(\boldsymbol{x}) + \cdots, \quad (2\text{-}3)$$

represents **an infinite series of functions**. Series (2-3) poses an additional question, that is, **for which values of x, if any, the series $\sum u_n(x)$ converges to a function $S(x)$?**

Expanding a given function (say $S(x)$), in an infinite series of other functions (say $u_k(x),\ k = 1,2,3,\cdots$), is a problem of great importance in Applied Mathematics, Engineering, Physics, etc.

Theorem 2-1.

If $|x| < 1$, then the infinite geometric series

$$\sum_{n=0}^{\infty} \boldsymbol{x^n} = \boldsymbol{1} + \boldsymbol{x^1} + \boldsymbol{x^2} + \boldsymbol{x^3} + \boldsymbol{x^4} + \cdots = \frac{\boldsymbol{1}}{\boldsymbol{1-x}} \quad (2\text{-}4)$$

Note: The general form of a geometric progression is $\{a, ax, ax^2, ax^3, ax^4, ax^5, \cdots\}$. **The number a, is called the first term of the progression, while x is the ratio of the progression**. If $|x| < 1$, the progression is called **decreasing**.

Proof: If we define

$S_n = 1 + x + x^2 + x^3 + \cdots + x^{n-1} + x^n$, then

$x \cdot S_n = x + x^2 + x^3 + x^4 + \cdots + x^n + x^{n+1}$,

and subtracting the second from the first, yields,

$$(1-x)\cdot S_n = 1 - x^{n+1} \Leftrightarrow S_n = \frac{1-x^{n+1}}{1-x}.$$

Taking the limit of both sides, and noting that

$\lim_{n\to\infty} x^{n+1} = 0$, (since $|x| < 1$), we have ,

$$\lim_{n\to\infty} S_n = \sum_{n=0}^{\infty} x^n = 1 + x + x^2 + x^3 + x^4 + \cdots = \frac{1}{1-x}, \qquad |x| < 1.$$

Note: The sum

$$\sum_{n=1}^{\infty} \boldsymbol{x^n} = \boldsymbol{x + x^2 + x^3 + x^4} + \cdots = \frac{\mathbf{1}}{\mathbf{1}-\boldsymbol{x}} - \mathbf{1} = \frac{\boldsymbol{x}}{\mathbf{1}-\boldsymbol{x}}, \quad |\boldsymbol{x}| < 1. \qquad \text{(2-5)}$$

Example 2-1.

Find the sum of the series $\sum_{n=0}^{\infty}(\frac{3}{2^n})$.

Solution

$$\sum_{n=0}^{\infty} \frac{3}{2^n} = 3\sum_{n=0}^{\infty} \frac{1}{2^n} = 3\sum_{n=0}^{\infty} \left(\frac{1}{2}\right)^n = 3 \cdot \frac{1}{1-\frac{1}{2}} = 6.$$

Example 2-2.

Find the sum of the series $\sum_{n=1}^{\infty}(\frac{(-1)^n}{5^n})$.

Solution

$$\sum_{n=1}^{\infty} \frac{(-1)^n}{5^n} = \sum_{n=1}^{\infty} \left(\frac{-1}{5}\right)^n = \frac{\frac{-1}{5}}{1-(-\frac{1}{5})} = -\frac{1}{6}.$$

(Application of (2-5) with $x = -\frac{1}{5}$, $|x| = \frac{1}{5} < 1$)

Example 2-3.

Does the series $\sum_{n=1}^{\infty}(-1)^n$ converge?

Solution

$$\sum_{n=1}^{\infty}(-1)^n = 1 - 1 + 1 - 1 + 1 - 1 + \cdots$$

The partial sums are,

$s_1 = 1,\ s_2 = 1 - 1 = 0,\ s_3 = 1 - 1 + 1 = 1,\ s_4 = 1 - 1 + 1 - 1 = 0, \ldots.$

i.e. the sequence of the partial sums is $(s_n) = \{1,0,1,0,1,0, \ldots.\}$, i.e. does not converge to any limit, therefore **the given series is not convergent**.

Example 2-4.

Find the rational number represented by the repeating decimal $0{,}353535\cdots$

Solution

Let $y = 0{,}353535\cdots = 0{,}35 + 0{,}0035 + 0{,}000035 + \cdots$, or

$$y = \frac{35}{10^2} + \frac{35}{10^4} + \frac{35}{10^6} + \cdots = 35 \cdot \sum_{n=1}^{\infty} \left(\frac{1}{10^2}\right)^n = 35 \cdot \frac{\frac{1}{10^2}}{1 - \frac{1}{10^2}} = \frac{35}{99}.$$

(Application of (2-5) with $x = \frac{1}{10^2} < 1$).

Another method:

If $y = 0{,}353535\cdots$, then $100y = 35{,}353535\cdots$, and subtracting the first from the second, $99y = 35 \Leftrightarrow y = \frac{35}{99}$.

Example 2-5.

Find the sum of the series $\sum_{n=1}^{\infty} \left(\frac{\cos\theta}{2}\right)^n$.

Solution

Since $|\cos\theta| \leq 1 < 2 \Rightarrow \left|\frac{\cos\theta}{2}\right| < 1$, and application of (2-5), with $x = \frac{\cos\theta}{2}$, yields,

$$\sum_{n=1}^{\infty} \left(\frac{\cos\theta}{2}\right)^n = \frac{\frac{\cos\theta}{2}}{1 - \frac{\cos\theta}{2}} = \frac{\cos\theta}{2 - \cos\theta}.$$

PROBLEMS

2-1) Find the sum of the following series,

a) $\sum_{n=2}^{\infty}\left(\frac{3}{4}\right)^n$, **b)** $\sum_{n=1}^{\infty}\left(-\frac{1}{3}\right)^n$, **c)** $\sum_{n=20}^{\infty}\left(\frac{1}{2}\right)^n$

(Answer: a) $\frac{9}{4}$, **b)** $-\frac{1}{4}$, **c)** $\frac{1}{2^{19}}$

2-2) If $x > 1$, find the sum of the series $\sum_{n=0}^{\infty}\frac{(-1)^n}{x^n}$.

Hint: Since $x > 1$, $0 < \frac{1}{x} < 1$.

2-3) Find a decreasing geometric progression, $(a, ax, ax^2, ax^3, ax^4, \cdots)$, such that the sum of its terms equals 4, while the sum of the cubes of its terms equals 192.

(Answer: $a = 6,\ x = -\frac{1}{2}$)

2-4) Find the sum of the following series,

a) $\sum_{n=0}^{\infty}\left(\frac{\sin\theta}{5}\right)^n$, **b)** $\sum_{n=1}^{\infty}\frac{1}{(x^2+5)^n}$

2-5) For which values of b, $\sum_{n=1}^{\infty}\frac{b^n}{(b+1)^{n-1}} < \infty$?

(Answer: $b > -\frac{1}{2}$)

2-6) Consider the finite sum,

$$S_n = 1 + 2x + 3x^2 + 4x^3 + \cdots + nx^{n-1}, \qquad (x \neq 1)$$

and show that,

$$S_n = \frac{n \cdot x^n}{x-1} - \frac{x^n - 1}{(x-1)^2} \cdot$$

2-7) If $|x| < 1$, show that $\sum_{n=0}^{\infty}(n+1)\cdot x^n = \frac{1}{(x-1)^2}$, and as an application sum the series $\sum_{n=0}^{\infty}\frac{n+1}{3^n}$.

(Answer: $\frac{9}{4}$)

Hint: Since $|x| < 1$, $\lim_{n\to\infty} x^n = 0$, $\lim_{n\to\infty}(nx^n) = 0$.

2-8) For which values of x, the series $\sum_{n=0}^{\infty}(-1)^n \frac{(1-x)^n}{(1+x)^{n+1}} < \infty$?

What is the sum of this series?

2-9) If b is the sum of the terms of an infinite decreasing progression, and c is the sum of the squares of its terms, find the sum of the first n terms of the progression.

(Answer: $b\left\{1 - \left(\frac{b^2-c}{b^2+c}\right)^n\right\} \cdot$)

2-10) Let $S_1 = \sum_{k=0}^{\infty} q^k$, ($|q| < 1$), and $S_2 = \sum_{k=0}^{\infty} p^k$, ($|p| < 1$). Express in terms of S_1 and S_2 the sum of the series $\sum_{k=0}^{\infty}(p \cdot q)^k$.

2-11) A rubber ball is dropped from an initial height $h = 10m$, above a flat surface. Each time the ball hits the surface, after falling a distance y, it rebounds a distance $\frac{5}{7}y$. Find the total distance the ball travels.

(Answer: $35m$)

3. Telescoping and Related Series.

In Chapter 2, we mentioned that in order to investigate the **nature of a series,** (i.e. to find whether the series is convergent or divergent), it suffices to investigate the sequence of the partial sums $(s_n) = \{s_1, s_2, s_3, \cdots, s_n, \cdots\}$.

Let us consider the series $\sum u_n$ having partial sums
$s_n = s_1 + s_2 + s_3 + \cdots s_n, \; n = 1,2,3,\cdots$

If the nth term s_n of the sequence (s_n) can be expressed as a function of n, i.e. if $s_n = f(n)$, then

$$\sum_{k=1}^{\infty} u_k = \lim_{n\to\infty}\left(\sum_{k=1}^{n} u_k\right) = \lim_{n\to\infty} s_n = \lim_{n\to\infty}(f(n)), \quad (3\text{-}1)$$

so **if $\lim_{n\to\infty}(f(n)) = S\ (finite)$, then $\sum u_k < \infty$,**

(i.e. the series converges),otherwise the series diverges.

Unfortunately, in very few cases, s_n can be expressed as a function of n. Some of these cases, the most important ones, are stated below:

Case I

Let us consider the series $\sum u_n$. If we can find a function $f(n)$, such that

$$u_n = f(n+1) - f(n), \quad n = 1, 2, 3, \cdots \quad (3\text{-}2)$$

then

$$s_n = u_1 + u_2 + u_3 + \cdots + u_n = (f(2) - f(1)) + (f(3) - f(2)) + (f(4) - f(3)) + \cdots + (f(n+1) - f(n))$$

or, $$s_n = f(n+1) - f(1), \quad (3\text{-}3)$$

and therefore,

$$\sum_{k=1}^{\infty} u_n = \lim s_n = \lim\{f(n+1) - f(1)\} = \lim f(n+1) - f(1). \quad (3\text{-}4)$$

The sum in (3-3) is **a telescoping sum**, meaning that each term cancels part of the next term, therefore the whole sum collapses into only two terms, (like a folding telescope).The corresponding series, is called **a telescoping series.**

If $\boldsymbol{u_n = f(n+2) - f(n)}$, $\boldsymbol{or}$ $\boldsymbol{u_n = f(n+3) - f(n)}$, etc., **where $\boldsymbol{f(n)}$ is a known function of $\boldsymbol{n}$**, the corresponding series will also be telescoping. For example, if

$u_n = f(n+2) - f(n)$, one may easily show that,

$$s_n = u_1 + u_2 + u_3 + \cdots + u_n = f(n+2) + f(n+1) - f(1) - f(2),$$

and therefore,

$$\boldsymbol{\sum_{n=1}^{\infty} u_n = \lim s_n = \lim\big(f(n+2) + f(n+1)\big) - f(1) - f(2).}$$

(See Example 3-1).

Case II

If the $n^{\underline{th}}$ term u_n of a series $\sum u_n$, can be expressed as

$$\boldsymbol{u_n = af(n) + bf(n+1) + cf(n+2),\ where\ (a+b+c) = 0}\,,$$

with $\boldsymbol{f(n)}$ **a known function of $\boldsymbol{n}$** , then s_n can be expressed, **in closed form,** as a function of n. The proof is easy and is shown below:

$$s_n = u_1 + u_2 + u_3 + \cdots + u_{n-1} + u_n =$$

$af(1) + bf(2) + cf(3) +$ (Term u_1)
$af(2) + bf(3) + cf(4) +$ (Term u_2)
$af(3) + bf(4) + cf(5) +$ (Term u_3)

....

....

....

$af(n-1) + bf(n) + cf(n+1) +$ (Term u_{n-1})
$af(n) + bf(n+1) + cf(n+2)$ (Term u_n)

Further simplification yields,

$$s_n = af(1) + (a+b)f(2) + (a+b+c)\{f(3) + f(4) + \cdots + f(n)\} + (b+c)f(n+1) + cf(n+2),$$

and since $(a+b+c) = 0$, by assumption,

$$s_n = af(1) - cf(2) - af(n+1) + cf(n+2),$$

i.e. s_n is finally expressed in terms of $f(n)$, in closed form, and therefore,

$\sum_{n=1}^{\infty} \boldsymbol{u_n} = \mathbf{lim}\, \boldsymbol{s_n} = \mathbf{lim}\{-\boldsymbol{af(n+1)} + \boldsymbol{cf(n+2)}\} + \boldsymbol{af(1)} - \boldsymbol{cf(2)}$. (3-5)

(See Examples 3-2 and 3-3).

Case III (Telescoping Trigonometric Series)

In many series, the general term u_n is a trigonometric function of n. Taking advantage of various Trigonometric Identities, it is possible sometimes (but not always) to express u_n as an appropriate difference, i.e. **find a function $\boldsymbol{f(n)}$ such that $\boldsymbol{u_n}$ can be expressed as**

$\boldsymbol{u_n} = \boldsymbol{f(n+1)} - \boldsymbol{f(n)}$.

We consider some characteristic cases, where the general term u_n and the corresponding function $f(n)$, are shown in the following Table:

1)	$u_n = \dfrac{1}{\cos\phi + \cos\big((2n+1)\phi\big)}$	$f(n) = \dfrac{\tan(n\phi)}{2\sin\phi}$	(3-6)
2)	$u_n = \cos(n\phi),\ \ \phi \neq 2k\pi$	$f(n) = \dfrac{\sin\left\{(2n-1)\frac{\phi}{2}\right\}}{2\sin(\frac{\phi}{2})}$	(3-7)
3)	$u_n = Arccot(n^2+n+1)$	$f(n) = Arctan(n)$	(3-8)
4)	$u_n = Arccot(2n^2)$	$f(n) = -Arccot(2n-1)$	(3-9)
5)	$u_n = \dfrac{\tan(2^n\phi)}{\cos(2^{n+1}\phi)}$	$f(n) = \tan(2^n\,\phi)$	(3-10)
6)	$u_n = 2^n\tan(\frac{\phi}{2^n})\left\{\tan(\frac{\phi}{2^{n+1}})\right\}^2$	$f(n) = -2^n\tan(\frac{\phi}{2^n})$	(3-11)

For a proof of (1) and (4), see Examples 3-4 and 3-5 respectively. For a proof of (2), (3), (5) and (6), see Problems 3-5, 3-6, 3-7 and 3-8 respectively.

(See Examples 3-6 and 3-7).

Example 3-1.

Find the sum of the series $\sum_{n=1}^{\infty}\frac{1}{n(n+1)}$.

Solution

The general term of the series is $u_n = \frac{1}{n(n+1)} = \frac{(n+1)-n}{n(n+1)} = \frac{1}{n} - \frac{1}{n+1}$, i.e.

$\boldsymbol{u_n = f(n) - f(n+1)}$, **where** $\boldsymbol{f(n) = \frac{1}{n}}$. Then,

$s_n = u_1 + u_2 + u_3 + \cdots u_n = (f(1) - f(2)) + (f(2) - f(3)) + (f(3) - f(4)) + \cdots + (f(n) - f(n+1)) = f(1) - f(n+1) = 1 - \frac{1}{n+1}$,

and therefore,

$$\sum_{n=1}^{\infty}\frac{1}{n(n+1)} = \lim s_n = \lim\left(1 - \frac{1}{n+1}\right) = 1 - 0 = 1.$$

The given series converges to the number 1.

Example 3-2.

Find the sum of the series $\sum_{n=1}^{\infty}\frac{1}{n(n+1)(n+2)}$.

Solution

The general term of the series is $u_n = \frac{1}{n(n+1)(n+2)}$.

Using **partial fraction decomposition,** u_n can be put in the form,

$$u_n = \frac{1}{n(n+1)(n+2)} = \frac{a}{n} + \frac{b}{n+1} + \frac{c}{n+2} \qquad (*)$$

where $\boldsymbol{a}$**,** $\boldsymbol{b}$ **and** $\boldsymbol{c}$ **are constants to be determined**. From (*) we have,

$$a(n+1)(n+2) + bn(n+2) + cn(n+1) = 1. \qquad (**)$$

Equation (**) is **an identity for** $\boldsymbol{n}$, i.e. it is true for all values of n. Applying (**) for $n = -1$, we have, $0 + b(-1)(-1+2) + 0 = 1$, or $\boldsymbol{b = -1}$, and similarly applying for $n = -2$ and $n = 0$, we find $\boldsymbol{c = \frac{1}{2}}$ and $\boldsymbol{a = \frac{1}{2}}$, respectively, therefore

$$u_n = \frac{1}{n(n+1)(n+2)} = \frac{1}{2}\cdot\frac{1}{n} - \frac{1}{n+1} + \frac{1}{2}\cdot\frac{1}{n+2}, \qquad (***)$$

and since $a + b + c = \frac{1}{2} - 1 + \frac{1}{2} = 0$, we are in Case II, and can apply (3-5) with $f(n) = \frac{1}{n}$ to obtain,

$$\sum_{n=1}^{\infty} \frac{1}{n(n+1)(n+2)} = \lim\left\{-\frac{1}{2(n+1)} + \frac{1}{2(n+2)}\right\} + \frac{1}{2}\cdot 1 - \frac{1}{2}\cdot\frac{1}{2} = \frac{1}{2} - \frac{1}{4} = \frac{1}{4}.$$

Example 3-3.

Find the sum of the series $\sum_{n=1}^{\infty} \frac{n^2+6n+12}{n(n+1)(n+2)} \cdot \frac{1}{2^{n+1}}$.

Solution

The general term of the series is $u_n = \frac{n^2+6n+12}{n(n+1)(n+2)} \frac{1}{2^{n+1}}$. The first factor in u_n can be **decomposed into partial fractions**, (since the degree of the numerator is $2 < 3$, the degree of the denominator), and working as in Example 3-2, we find,

$$\frac{n^2+6n+12}{n(n+1)(n+2)} = \frac{6}{n} - \frac{7}{n+1} + \frac{2}{n+2}, \qquad (*)$$

(for a proof see Problem 3-21), and finally the general term u_n can be written as,

$$u_n = \frac{n^2+6n+12}{n(n+1)(n+2)} \cdot \frac{1}{2^{n+1}} = \frac{3}{n2^n} - \frac{7}{(n+1)2^{n+1}} + \frac{4}{(n+2)2^{n+2}}.$$

We are in Case II, with $f(n) = \frac{1}{n\cdot 2^n}$, $a = 3,\ b = -7,\ c = 4,\ (a + b + c = 0)$, and application of (3-5) yields,

$$\sum_{n=1}^{\infty} \frac{n^2+6n+12}{n(n+1)(n+2)} \cdot \frac{1}{2^{n+1}} = \lim\left\{-\frac{3}{(n+1)2^{n+1}} + \frac{4}{(n+2)2^{n+2}}\right\} + \frac{3}{1\cdot 2} - 4\frac{1}{2\cdot 2^2} = 1.$$

Example 3-4.

Prove the Trigonometric identity (3-6).

Solution

It suffices to show that

$$f(n+1)-f(n)=\frac{\tan((n+1)\phi)}{2\sin\phi}-\frac{\tan(n\phi)}{2\sin\phi}=\frac{1}{\cos\phi+\cos((2n+1)\phi)}=u_n.$$

We have,

$$\frac{\tan((n+1)\phi)}{2\sin\phi}-\frac{\tan(n\phi)}{2\sin\phi}=\frac{1}{2\sin\phi}\left\{\frac{\sin((n+1)\phi)}{\cos((n+1)\phi)}-\frac{\sin(n\phi)}{\cos(n\phi)}\right\}$$
$$=\frac{\sin((n+1)\phi)\cos(n\phi)-\sin(n\phi)\cos((n+1)\phi)}{2\sin\phi\cos((n+1)\phi)\cos(n\phi)}$$
$$=\frac{\sin((n+1)\phi-n\phi)}{\sin\phi\{\cos((n+1)\phi+n\phi)+\cos((n+1)\phi-n\phi)\}}$$
$$=\frac{\sin\phi}{\sin\phi\{\cos((2n+1)\phi)+\cos\phi\}}=\frac{1}{\cos((2n+1)\phi)+\cos\phi},$$

and the proof is thus completed.

Note: In the derivation above we have used two well known Trigonometric Identities, i.e. $\mathbf{\sin x\cos y-\sin y\cos x=\sin(x-y)}$, and $\mathbf{2\cos x\cos y=\cos(x+y)+\cos(x-y)}$.

Example 3-5.

Prove formula (3-9).

Solution

We shall show first the following Trigonometric Identity:

If $\mathbf{y>x>0}$**, then Arc cot**x **−Arc cot**y **=Arc cot**$(\frac{1+xy}{y-x})$ **,** (*)

where by **Arc cot we mean the principal branch of the inverse cotangent function**, meaning that ,if

$$a=Arccotx\Leftrightarrow cota=x,\quad and\quad 0<a<\pi. \qquad (**)$$

Let $a=Arccotx\Leftrightarrow cota=x,\quad 0<a<\frac{\pi}{2},$ (since $x>0$), (***)

$b=Arccoty\Leftrightarrow cotb=y,\quad 0<b<\frac{\pi}{2},$ (since $y>0$), (****)

$c=Arccot(\frac{1+xy}{y-x})\Leftrightarrow cotc=\frac{1+xy}{y-x},\quad 0<c<\frac{\pi}{2}$ (since $\frac{1+xy}{y-x}>0$). (*****)

We want to show that $\boldsymbol{a - b = c}$**.**

We note that $\cot(a-b) = \frac{1+cota\cdot cotb}{cotb-cota} = \frac{1+xy}{y-x} = cotc$, from which we conclude that

$a - b = c + k\pi \Leftrightarrow \boldsymbol{a - b - c = k\pi}$, where $k = 0, \pm 1, \pm 2, \pm 3, \cdots$

In order therefore to show that $a - b = c$, **it suffices to show that** $\boldsymbol{k = 0}$**.**

Indeed, from the inequalities (***), (****), and (*****), we have,

$-\pi < a - b - c < \frac{\pi}{2} \Leftrightarrow -\pi < k\pi < \frac{\pi}{2} \Leftrightarrow -1 < k < \frac{1}{2} \Rightarrow \boldsymbol{k = 0},$

since the only integer between (-1) and $(\frac{1}{2})$ is the number 0, and the proof is therefore completed.

If we apply now (*), for $x = 2n - 1$, and $y = 2n + 1$, we obtain,

$$Arccot(2n-1) - Arccot(2n+1) = Arccot\frac{1+(2n-1)\cdot(2n+1)}{(2n+1)-(2n-1)} = Arccot(2n^2),$$

and the proof of (3-9) follows immediately.

Example 3-6.

Evaluate the finite sum $S_k = \sum_{n=1}^{k} Arccot(2n^2)$, and then find the sum of the infinite series, $\sum_{n=1}^{\infty} Arccot(2n^2)$.

Solution

From (3-9), $Arccot(2n^2) = -Arccot(2n+1) + Arccot(2n-1)$. (*)

Applying (*) for $n = 1,2,3,\cdots,k$ and adding term wise, we obtain,

$$S_k = \sum_{n=1}^{k} Arccot(2n^2) = Arccot1 - Arccot(2k+1) =$$

$$= \frac{\pi}{4} - Arccot(2k+1).$$

Passing to the limit as $k \to \infty$, we have,

$$\lim_{k\to\infty} S_k = \sum_{n=1}^{\infty} Arccot(2n^2) =$$

$$= \frac{\pi}{4} - \lim_{k\to\infty} Arccot(2k+1) = \frac{\pi}{4} - Arccot(+\infty) = \frac{\pi}{4} - 0 = \frac{\pi}{4}.$$

Example 3-7.

Making use of the Trigonometric Identity $\tan(\frac{x}{2}) = \cot(\frac{x}{2}) - 2\cot x$, (for a proof see Problem 3-2), find the finite sum $S_n = \sum_{k=1}^{n} \frac{1}{2^k}\tan(\frac{x}{2^k})$, and then pass to the limit as $n \to \infty$, to evaluate the sum of the infinite series $\sum_{k=1}^{\infty} \frac{1}{2^k}\tan(\frac{x}{2^k})$.

Solution

From the given Trigonometric Identity, we have,

$$\frac{1}{2^k}\tan(\frac{x}{2^k}) = \frac{1}{2^k}\cot(\frac{x}{2^k}) - \frac{1}{2^{k-1}}\cot(\frac{x}{2^{k-1}}) = f(k) - f(k-1),$$

where $\boldsymbol{f(k) = \frac{1}{2^k}\cot(\frac{x}{2^k})}$.

Applying the equation above, for $k = 1,2,3,\cdots,n$ and adding term wise, we have,

$$S_n = \sum_{k=1}^{n} \frac{1}{2^k}\tan\left(\frac{x}{2^k}\right) =$$

$$= (f(1) - f(0)) + (f(2) - f(1)) + \cdots + (f(n) - f(n-1))$$
$$= f(n) - f(0),$$

and finally,

$$\boldsymbol{S_n = \sum_{k=1}^{n} \frac{1}{2^k} \cdot \tan\left(\frac{x}{2^k}\right) = \frac{1}{2^n}\cot(\frac{x}{2^n}) - \cot x.}$$

The infinite series

$$\sum_{n=1}^{\infty} \frac{1}{2^n}\tan(\frac{x}{2^n}) = \lim_{n\to\infty} S_n = \lim_{n\to\infty}\left(\frac{1}{2^n}\cot\left(\frac{x}{2^n}\right) - \cot x\right)$$

$$= \lim_{n\to\infty}\left(\frac{1}{2^n}\cot\left(\frac{x}{2^n}\right)\right) - \cot x$$

$$= \lim_{n\to\infty}\left\{\frac{1}{x} \cdot \frac{\frac{x}{2^n}}{\sin(\frac{x}{2^n})} \cdot \cos(\frac{x}{2^n})\right\} - \cot x.$$

As $n \to \infty$, $\frac{x}{2^n} \to 0$, $\frac{(x/_{2^n})}{\sin(x/_{2^n})} \to 1$, (see Note), while $\cos(x/_{2^n}) \to 1$, and finally,

$$\sum_{n=1}^{\infty} \frac{\mathbf{1}}{\mathbf{2^n}} \mathbf{\tan}\left(\frac{\boldsymbol{x}}{\mathbf{2^n}}\right) = \frac{\mathbf{1}}{\boldsymbol{x}} \cdot \mathbf{1} \cdot \mathbf{1} - \mathbf{\cot}\, \boldsymbol{x} = \frac{\mathbf{1}}{\boldsymbol{x}} - \mathbf{\cot}\, \boldsymbol{x}.$$

Note: If a variable quantity $y \to 0$, then $\lim_{y \to 0} \cos y = 1$, while $\lim_{y \to 0} \frac{\sin y}{y} = 1 = \lim_{y \to 0} \frac{y}{\sin y}$. Also $\lim_{y \to 0} \frac{\tan y}{y} = \lim_{y \to 0} \left(\frac{\sin y}{y} \cdot \frac{1}{\cos y}\right) = \lim_{y \to 0} \frac{\sin y}{y} \cdot \lim_{y \to 0} \frac{1}{\cos y} = 1.$

Example 3-8.

Show the identity $\boldsymbol{n^3(n+1)^3 - (n-1)^3 n^3 = 6n^5 + 2n^3}$ and then evaluate the sum $S = 1^5 + 2^5 + 3^5 + \cdots k^5$.

Solution

The expression $n^3(n+1)^3 - (n-1)^3 n^3 = n^3\{(n+1)^3 - (n-1)^3\} = n^3\{(n^3 + 3n^2 + 3n + 1) - (n^3 - 3n^2 + 3n - 1)\} = n^3(6n^2 + 2) = 6n^5 + 2n^3$, or

$6n^5 + 2n^3 = f(n) - f(n-1)$ where $f(n) = n^3(n+1)^3$.

Applying this formula for $n = 1,2,3, \cdots, k-1, k$ and adding the derived equalities term wise, we obtain,

$6(1^5 + 2^5 + 3^5 + \cdots + (k-1)^5 + k^5) + 2(1^3 + 2^3 + 3^3 + \cdots + (k-1)^3 + k^3) = \big(f(1) - f(0)\big) + \big(f(2) - f(1)\big) + \big(f(3) - f(2)\big) + \cdots + \big(f(k-1) - f(k-2)\big) + \big(f(k) - f(k-1)\big) = f(k) - f(0) = k^3(k+1)^3 - 0 = k^3(k+1)^3$, or

$6S + 2\left(\frac{k(k+1)}{2}\right)^2 = k^3(k+1)^3$, by virtue of Problem 1-3, from which

$\boldsymbol{S = \frac{k^2(k+1)^2(2k^2+2k-1)}{12}}$.

PROBLEMS

3-1) Find the sum of the series $\frac{1}{1 \cdot 2 \cdot 3 \cdot 4} + \frac{1}{2 \cdot 3 \cdot 4 \cdot 5} + \frac{1}{3 \cdot 4 \cdot 5 \cdot 6} + \cdots$

(**Answer:** $\frac{1}{18}$).

3-2) Show the trigonometric identity, $\tan(\frac{x}{2}) = \cot(\frac{x}{2}) - 2\cot x$.

3-3) Find the sum of the series $\frac{1}{1\cdot 3} + \frac{1}{2\cdot 4} + \frac{1}{3\cdot 5} + \cdots$

(Answer: $\frac{3}{4}$).

3-4) Find the sum of the series $\sum_{n=1}^{\infty} \frac{n+2}{n(n+1)} \cdot \frac{1}{2^n}$.

3-5) Prove formula (3-7).

3-6) Prove formula (3-8).

3-7) Prove formula (3-10).

3-8) Prove formula (3-11).

3-9) Sum the series $\sum_{n=1}^{\infty} \frac{2n-1}{n(n+1)(n+3)}$

(Answer: $\frac{23}{36}$).

3-10) Sum the series $\sum_{n=1}^{\infty} \frac{1+3^n+7^n}{8^n}$

3-11) Sum the series $\sum_{n=1}^{\infty} \frac{1}{n\sqrt{n+1}+(n+1)\sqrt{n}}$

Hint: $\frac{1}{n\sqrt{n+1}+(n+1)\sqrt{n}} = \frac{n\sqrt{n+1}-(n+1)\sqrt{n}}{n^2(n+1)-(n+1)^2 n} = \frac{1}{\sqrt{n}} - \frac{1}{\sqrt{n+1}}$, etc.

(Answer: 1).

3-12) Sum the series $\sum_{n=1}^{\infty} \frac{1}{(3n-1)(3n+2)(3n+5)}$

3-13) Sum the series $\sum_{k=1}^{\infty} 2^k \tan\left(\frac{x}{2^k}\right) (\tan(\frac{x}{2^{k+1}}))^2$

Hint: Make use identity (3-11).

(Answer: $2\tan\left(\frac{x}{2}\right) - \frac{1}{x}$).

3-14) Evaluate the finite sum $\sum_{m=1}^{n} \frac{1}{\cos x+\cos\{(2m+1)x\}}$

Hint: Make use of identity (3-6).

3-15) Show that $\sum_{k=1}^{\infty} \frac{k}{1\cdot 3\cdot 5\cdots(2k+1)} = \frac{1}{2}$

Hint: The general term of the given series is ,

$$u_k = \frac{k}{1\cdot3\cdot5\cdots(2k+1)} = \frac{1}{2}\cdot\frac{(2k+1)-1}{1\cdot3\cdot5\cdots(2k+1)} = \frac{1}{2}\left\{\frac{1}{1\cdot3\cdot5\cdots(2k-1)} - \frac{1}{1\cdot3\cdot5\cdots(2k+1)}\right\}, \quad or$$

$u_k = \frac{1}{2}(f(k) - f(k+1))$, where $\boldsymbol{f(k)} = \frac{\mathbf{1}}{\mathbf{1\cdot3\cdot5\cdots(2k-1)}}$, etc.

3-16) Sum the series $\sum_{k=1}^{\infty}\frac{k\cdot2^k}{(k+2)!}$

Hint: The general term of the series is

$$u_k = \frac{k2^k}{(k+2)!} = \frac{\{(k+2)-2\}2^k}{(k+2)!} = \frac{2^k}{(k+1)!} - \frac{2^{k+1}}{(k+2)!} \text{ , etc.}$$

3-17) Show that $\sum_{m=1}^{\infty}\frac{m^2+m-1}{(m+2)!} = \frac{1}{2}$

Hint: $m^2 + m - 1 = m^2 + 2m - m - 1 = m(m+2) - (m+1)$, etc.

3-18) If $u_n = \frac{n^2+6n+12}{n^3+3n^2+2n}\cdot\frac{1}{2^{n+1}}$, evaluate $\sum_{n=1}^{\infty}u_n$.

Hint: Factorize the denominator and apply partial fractions decomposition.

3-19) Evaluate the sum $\sum_{n=1}^{\infty}\frac{n}{(n+2)(n+3)(n+4)}$

(Answer: $\frac{1}{6}$).

3-20) Find the sum of the series $\sum_{n=1}^{\infty}\frac{1}{n(n+1)(n+2)\cdots(n+10)}$

3-21) Prove equation (*) in Example 3-3.

3-22) Show that $\sum_{n=0}^{\infty}\frac{n+1}{2^n} = 4.$

Hint: If $S = \sum_{n=0}^{\infty}\frac{n+1}{2^n}$ then $\frac{1}{2^2}S = \sum_{n=0}^{\infty}\frac{n+1}{2^{n+2}}$, and subtracting term wise we have, $S\cdot\left(1-\frac{1}{2^2}\right) = 1 + 1 + \frac{2}{2^2}\left(1 + \frac{1}{2} + \frac{1}{2^2} + \frac{1}{2^3} + \cdots\right)$, etc

3-23) Sum the series $\sum_{n=1}^{\infty}\frac{3n^2+3n+1}{n^3(n+1)^3}$

(Answer: 1)

Hint: The term $u_n = \frac{3n^2+3n+1}{n^3(n+1)^3} = \frac{(n+1)^3-n^3}{n^3(n+1)^3} = \frac{1}{n^3} - \frac{1}{(n+1)^3}$

3-24) Show that

a) $\sum_{n=1}^{\infty} \frac{\sqrt{n+1}-\sqrt{n}}{\sqrt{n^2+n}} = 1$ and **b)** $\sum_{n=1}^{\infty} \frac{2^n+3^n}{6^n} = \frac{3}{2}$

3-25) Sum the series $\sum_{n=1}^{\infty} Arctan(\frac{2}{n^2})$.

Hint: Show that $\boldsymbol{Arctan(n+1) - Arctan(n-1) = Arctan(\frac{2}{n^2})}$, (see Example 3-5).

(**Answer:** $\frac{3\pi}{4}$).

3-26) If $Arcsinx$ is **the principal branch of the inverse sine function,** i.e. $\boldsymbol{y = Arcsinx \Leftrightarrow siny = x}$**, and** $\boldsymbol{-1 \le x \le 1}$**, and** $\boldsymbol{-\frac{\pi}{2} \le y \le \frac{\pi}{2}}$,
show that

$$Arcsin\frac{\sqrt{(k+1)^2-1}-\sqrt{k^2-1}}{k(k+1)} = Arcsin\left(\frac{1}{k}\right) - Arcsin\left(\frac{1}{k+1}\right),$$

where k is any positive integer, and then show that

$$\sum_{k=1}^{\infty} Arcsin\frac{\sqrt{(k+1)^2-1}-\sqrt{k^2-1}}{k(k+1)} = \frac{\pi}{2}$$

3-27) Show the identity $\boldsymbol{n^2(n+1)^2(2n+1) - (n-1)^2n^2(2n-1) = 10n^4 + 2n^2}$ and then find the sum $S = 1^4 + 2^4 + 3^4 + \cdots + k^4$ in closed form.

Hint: Make use of Problem 1-2.

(**Answer:** $S = \frac{k(k+1)(2k+1)(3k^2+3k-1)}{30}$).

3-28) Show the identity $\boldsymbol{n^3(n+1)^3(2n+1) - (n-1)^3n^3(2n-1) = 14n^6 + 10n^4}$ and then find the sum $S = \sum_{n=1}^{k} n^6$.

4. General Theorems on Series.

In this Chapter we will develop some general Theorems on series, by means of which, one may rather easily **investigate whether a given series is convergent or not.**

Theorem 4-1.

If $\sum_{k=1}^{\infty} u_k$ and $\sum_{k=1}^{\infty} \phi_k$ are convergent series, then the series

$\sum_{k=1}^{\infty}(u_k + \phi_k)$ and $\sum_{k=1}^{\infty}(u_k - \phi_k)$ are also convergent and furthermore,

$\sum_{k=1}^{\infty}(u_k + \phi_k) = \sum_{k=1}^{\infty} u_k + \sum_{k=1}^{\infty} \phi_k$ and (4-1)

$\sum_{k=1}^{\infty}(u_k - \phi_k) = \sum_{k=1}^{\infty} u_k - \sum_{k=1}^{\infty} \phi_k$. (4-2)

Also if c is a nonzero constant, then the series $\sum_{k=1}^{\infty}(cu_k)$ is convergent, and

$\sum_{k=1}^{\infty}(cu_k) = c\sum_{k=1}^{\infty} u_k$. (4-3)

Proof: Let (s_n) and (t_n) be the sequences of the partial sums of $\sum_{k=1}^{\infty} u_k$ and $\sum_{k=1}^{\infty} \phi_k$ respectively, i.e.

$s_n = u_1 + u_2 + u_3 + \cdots u_n$ and $t_n = \phi_1 + \phi_2 + \phi_3 + \cdots \phi_n$. (*)

Since by assumption, the infinite series $\sum_{k=1}^{\infty} u_k$ and $\sum_{k=1}^{\infty} \phi_k$ are convergent, say to the numbers a and b respectively, the $\mathbf{\lim s_n = a}$ and $\mathbf{\lim t_n = b}$, therefore

$$\lim(s_n + t_n) = \lim s_n + \lim t_n = a + b \Rightarrow \sum_{k=1}^{\infty}(u_k + \phi_k) =$$

$$= \sum_{k=1}^{\infty} u_k + \sum_{k=1}^{\infty} \phi_k = a + b$$

and this completes the proof of (4-1).

Quite similarly, one may prove (4-2) and (4-3).

Note: If $\sum_{k=1}^{\infty} u_k < \infty$, $\sum_{k=1}^{\infty} \phi_k < \infty$, **and c and d are any two nonzero constants**, then $\sum_{k=1}^{\infty}(cu_k + d\phi_k) = c\sum_{k=1}^{\infty} u_k + d\sum_{k=1}^{\infty} \phi_k$. (4-4)

Theorem 4-2.

The convergence or the divergence of a series is unaffected by deleting a finite number of terms from the beginning of the series or by inserting a finite number of terms at the beginning of the series.

For Example, let us assume that a series $\sum_{k=1}^{\infty} u_k$ converges to a finite number ℓ, i.e. $\boldsymbol{u_1 + u_2 + u_3 + \cdots + u_n + \cdots = \ell}$.

If we **delete the first two terms**, the resulting series
$u_3 + \cdots + u_n + \cdots = \ell - (u_1 + u_2)$, is also convergent, i.e.
$\sum_{k=3}^{\infty} u_k = \sum_{k=1}^{\infty} u_{k+2} = \ell - (u_1 + u_2) < \infty$.

If we **insert, say three new terms $\boldsymbol{b_1, b_2}$, and $\boldsymbol{b_3}$** at the beginning of the series, the new series
$b_1 + b_2 + b_3 + u_1 + u_2 + u_3 + \cdots + u_n + \cdots = (b_1 + b_2 + b_3) + \ell < \infty$,
meaning that this new series is also convergent.

We say that two series are **of the same nature, if both are convergent or both are divergent.** In that sense, Theorem 4-2 states that **deleting or inserting a finite number of terms, does not alter the nature of the series.**

It does however alter the sum of the series.

For Example the series

$1 + \frac{1}{3} + \frac{1}{3^2} + \frac{1}{3^3} + \cdots = \frac{1}{1 - \frac{1}{3}} = \frac{3}{2}$, (a decreasing geometric progression), while

$\frac{1}{3^2} + \frac{1}{3^3} + \frac{1}{3^4} \cdots = \frac{3}{2} - 1 - \frac{1}{3} = \frac{1}{6} \neq \frac{3}{2}$ and

$2 + 3 + 4 + 1 + \frac{1}{3} + \frac{1}{3^2} + \frac{1}{3^3} = 9 + \frac{3}{2} = \frac{21}{2} \neq \frac{3}{2}$.

Proof: Let $\boldsymbol{\sum_{k=1}^{\infty} u_k = u < \infty}$. The sequence of partial sums is

$\{s_1 = u_1,\ s_2 = u_1 + u_2,\ s_3 = u_1 + u_2 + u_3,\ \cdots\}$

If from the given series **delete the first $\boldsymbol{(k-1)}$ terms**, where
$\boldsymbol{u_1 + u_2 + + \cdots + u_{k-1} = A}$, the resulting new series will be,
$u_k + u_{k+1} + u_{k+2} + \cdots$. The sequence of partial sums, for this new series, is

$\{s'_1 = u_k,\ s'_2 = u_k + u_{k+1},\ s'_3 = u_k + u_{k+1} + u_{k+2},\ \cdots\}$, or equivalently,

$\{s'_1 = s_k - A,\ s'_2 = s_{k+1} - A,\ s'_3 = s_{k+2} - A,\ \cdots,\ s'_n = s_{k+n-1} - A,\ \cdots\}.$

Since **by assumption the sequence (s_n) converges,** so does the sequence $(s_n - A)$, i.e. the sequence (s'_n), and $\lim(s'_n) = \lim(s_n - A) \Rightarrow \sum_{n=k}^{\infty} u_n = \sum_{n=1}^{\infty} u_n - A < \infty$, meaning that **the series $u_k + u_{k+1} + u_{k+2} + \cdots$ is convergent, (same nature with $u_1 + u_2 + u_3 + \cdots$) and its sum is equal to $(u - A)$.**

If $\sum_{k=1}^{\infty} u_k$ is divergent, then the sequences (s_n) and $(s'_n = s_n - A)$ do not converge, i.e. the series $u_k + u_{k+1} + u_{k+2} + \cdots$ does not converge to a finite limit, it is a divergent series, like the original series $u_1 + u_2 + u_3 + \cdots$.

In a similar manner we may show that inserting a finite number of terms, does not alter the nature of the original series.

Note: More generally, it can be shown that **inserting, deleting or even altering a finite number of terms of an infinite series, does not affect the nature of the series. It does alter, however, the sum of the original series, in case the original series is convergent.**

Theorem 4-3.

Grouping the terms of a convergent series, in an arbitrary manner, does not alter the nature and the sum of the original series.

For Example if $u_1 + u_2 + u_3 + u_4 + u_5 + u_6 + u_7 + u_8 + \cdots = u,$

then, $u_1 + (u_2 + u_3 + u_4) + (u_5 + u_6) + (u_7 + u_8) + \cdots = u,$

or, $(u_1 + u_2 + u_3) + (u_4 + u_5) + (u_6 + u_7 + u_8) + \cdots = u,$ etc.

Proof: Let $\sum_{n=1}^{\infty} u_n = u < \infty$, be a convergent series, and let us define a new series $\phi_1 + \phi_2 + \phi_3 + \cdots$, as follows,

$$\underbrace{(u_1 + u_2)}_{\phi_1} + \underbrace{(u_3 + u_4 + u_5)}_{\phi_2} + \underbrace{(u_6)}_{\phi_3} + \cdots + \underbrace{(u_r + u_{r+1} + \cdots + u_{r+k})}_{\phi_m} + \cdots, \ (*)$$

If (s_n) is the sequence of the partial sums of the original series, i.e.,

$\{s_1 = u_1,\ s_2 = u_1 + u_2,\ s_3 = u_1 + u_2 + u_3, \cdots\}$, and (w_n) is the sequence of partial sums of the new series, i.e.

$\{w_1 = \phi_1,\ w_2 = \phi_1 + \phi_2,\ w_3 = \phi_1 + \phi_2 + \phi_3,\ \cdots\}$, we see that the **sequence (w_n) is a subsequence of (s_n)**, i.e.
$\{w_1 = \phi_1 = s_2, w_2 = \phi_1 + \phi_2 = s_5, w_3 = \phi_1 + \phi_2 + \phi_3 = s_6, \cdots\}$,
and since the $\lim s_n = u < \infty$, **its subsequence (w_n) tends to the same limit u**, i.e.

$\lim w_n = \sum_{n=1}^{\infty} \phi_n = u = \sum_{n=1}^{\infty} u_n$, and this completes the proof.

Note: Theorem 4-3 actually states that we can arbitrarily group the terms of a convergent series, (by introducing brackets),and this has no effect on the nature and the sum of the original series.

A natural question arises at this point.

Can we always remove existing brackets in a convergent series? The answer, in general, is negative.

For instance, let us consider the series,

$$(1-1)+(1-1)+(1-1)+(1-1)+\cdots = 0+0+0+0+\cdots = 0.$$

If we remove the parentheses we have, $1-1+1+1+1-1+1-1+\cdots$, and this series **has no limit**, (see Example 2-3).

Before the difference between convergent and divergent series was understood, **and rigorously defined by Cauchy**, in terms of the limiting procedure involving the sequence of the partial sums of the series, this fact was consider as a paradox. Today it is not.

Note: There is a special case, **for series with positive terms, where removing brackets in a convergent series, does not alter its sum**. (For a proof see problem 4-1).

Theorem 4-4.

(Cauchy's necessary and sufficient condition for the convergence of a series, with arbitrary terms).

An infinite series $\sum_{n=1}^{\infty} u_n$ converges, if and only if, for every $\varepsilon > 0$, (arbitrarily small positive number ε) there exists a positive integer $N = N(\varepsilon)$, such that , for all the pair of indices n and m, with $m > n \geq N$, to have,

$$|\boldsymbol{u}_{n+1} + \boldsymbol{u}_{n+2} + \boldsymbol{u}_{n+3} + \cdots + \boldsymbol{u}_m| < \varepsilon. \quad (4\text{-}5)$$

This simply means that from a certain point on, **any block of consecutive terms, no matter how long**, will have a very small sum, (smaller that any arbitrarily small $\varepsilon > 0$).

Proof: The proof of Theorem 4-4 is an immediate consequence of the **Cauchy's necessary and sufficient condition for the convergence of a sequence**. Indeed, let us consider the series $\sum_{n=1}^{\infty} u_n$, and the sequence (s_n) of its partial sums, i.e.

$$\{s_1 = u_1,\ s_2 = u_1 + u_2,\ s_3 = u_1 + u_2 + u_3,\ \cdots\}. \quad (*)$$

According to the Cauchy's general convergence criterion, the sequence (s_n) will be convergent, (and therefore the series $\sum_{n=1}^{\infty} u_n$ will be convergent as well), **if and only if,**

$$\forall \varepsilon > 0 \quad \exists N = N(\varepsilon)\colon\ \forall m > n \geq N \Rightarrow |s_m - s_n| < \varepsilon,$$

or equivalently,

$$|\sum_{k=1}^{m} \boldsymbol{u}_k - \sum_{k=1}^{n} \boldsymbol{u}_k| < \varepsilon \Leftrightarrow |\boldsymbol{u}_{n+1} + \boldsymbol{u}_{n+2} + \boldsymbol{u}_{n+3} + \cdots + \boldsymbol{u}_m| < \varepsilon,$$

and this completes the proof.

Theorem 4-5 (Necessary condition for convergence).

If a series $\sum_{n=1}^{\infty} u_n$ converges, then its general term u_n tends to zero, as $n \to \infty$. In symbols,

$$\textbf{If } \sum_{n=1}^{\infty} \boldsymbol{u}_n = \boldsymbol{\ell} < \infty, \textbf{ then } \lim \boldsymbol{u}_n = \mathbf{0}. \quad (4\text{-}6)$$

Proof: If (s_n) is the sequence of partial sums of the convergent series $\sum_{n=1}^{\infty} u_n$, then

$$u_n = (u_1 + u_2 + u_3 + \cdots + u_{n-1} + u_n) - (u_1 + u_2 + u_3 + \cdots + u_{n-1}) =$$

$$= s_n - s_{n-1},$$

and passing to the limit as $n \to \infty$, we have

$$\lim u_n = \lim s_n - \lim s_{n-1} = \ell - \ell = 0,$$

and this completes the proof.

Note: Misuse of this Theorem may cause painful errors. **Theorem 4-5 is actually a Test for divergence and not a Test for convergence**. It simply states that,

If $\lim u_n \neq 0$, then the series $\sum_{n=1}^{\infty} u_n$ diverges.

However, **if the $\lim u_n = 0$, the series $\sum_{n=1}^{\infty} u_n$ may diverge or it may converge.**

As an Example let us consider the two series, $\sum_{n=1}^{\infty} \frac{1}{n}$ and $\sum_{n=1}^{\infty} \frac{1}{n^2}$.The general term in the first series is $u_n = \frac{1}{n}$, while in the second series is $w_n = \frac{1}{n^2}$.

In both cases,

$\lim u_n = \lim \frac{1}{n} = 0$, and $\lim w_n = \lim \frac{1}{n^2} = 0$. However, as we shall show shortly, the first series diverges to $+\infty$, while the second series converges to the finite number $\frac{\pi^2}{6}$.i.e.

$\sum_{n1}^{\infty} \frac{1}{n} = +\infty$, and $\sum_{n=1}^{\infty} \frac{1}{n^2} = \frac{\pi^2}{6}$.

Theorem 4-6 (The Harmonic Series).

The Harmonic series $\sum_{n=1}^{\infty} \frac{1}{n} = \frac{1}{1} + \frac{1}{2} + \frac{1}{3} + \cdots = +\infty$.

Proof: The series $\sum_{n=1}^{\infty} \frac{1}{n} = \frac{1}{1} + \frac{1}{2} + \frac{1}{3} + \cdots = +\infty$ is known as the **harmonic series,** since its terms constitute an harmonic progression. (We say that the numbers $b_1, b_2, b_3, \cdots$ form an **harmonic progression** if their reciprocals $\frac{1}{b_1}, \frac{1}{b_2}, \frac{1}{b_3}, \cdots$ form an **arithmetic progression**). Let us consider the sum

$$u_{n+1} + u_{n+2} + u_{n+3} + \cdots + u_{2n} = \frac{1}{n+1} + \frac{1}{n+2} + \frac{1}{n+3} + \cdots + \frac{1}{2n} >$$

$$\frac{1}{2n} + \frac{1}{2n} + \frac{1}{2n} + \cdots + \frac{1}{2n} = \frac{1}{2}$$

i.e. $|u_{n+1} + u_{n+2} + u_{n+3} + \cdots + u_{2n}| > \frac{1}{2}$, meaning that if we choose $m = 2n$ and $0 < \varepsilon < \frac{1}{2}$, (4-5) is not satisfied, therefore **the series $\sum_{n=1}^{\infty} \frac{1}{n}$ does not converge to a finite limit.** This in turn means that the sequence of partial sums (s_n) is an increasing sequence of positive numbers, $(s_1 > 0, s_2 > s_1 > 0, s_3 > s_2 > 0, \cdots)$, without having an upper bound, (because if it did, it would converge to a limit, but as we have just proved such a limit does not exist), therefore the sequence of partial sums (s_n) tends to $+\infty$, i.e.

$$\lim s_n = +\infty \Rightarrow \sum_{n=1}^{\infty} \frac{1}{n} = +\infty,$$

and this completes the proof.

Note: The harmonic series is a classic example of a series, such that $\lim u_n = \lim \left(\frac{1}{n}\right) = 0$, but the series itself diverges.

The French mathematician **Nicole Oresme (1320-1382)** seems to be the first one who proved that the harmonic series diverges.

Example 4-1.

Show that the series $\sum_{n=1}^{\infty} \frac{n(n+1)}{2n^2+5n+10}$ diverges.

Solution

The general term of the given series is $u_n = \frac{n(n+1)}{2n^2+5n+10}$ with $\lim u_n = \frac{1}{2} \neq 0$, and therefore the series diverges, (Theorem 4-5).

Example 4-2.

Is the series $\sum_{n=1}^{\infty} \frac{1}{\sqrt[n]{n}}$ convergent or divergent?

Solution

The general term of the given series is $u_n = \frac{1}{\sqrt[n]{n}}$ and since $\lim \sqrt[n]{n} = 1$, the $\lim u_n = \lim \left(\frac{1}{\sqrt[n]{n}}\right) = \frac{1}{1} \neq 0$, therefore the given series diverges, i.e. $\sum_{n=1}^{\infty} \frac{1}{\sqrt[n]{n}} = +\infty$.

Example 4-3.

In this Example we will demonstrate that **rearranging the terms of a convergent series, in general, the sum of the series is altered.**

Let us consider the series $\sum_{n=1}^{\infty}\frac{(-1)^{n-1}}{n}$ which is known to be convergent to the number $\ln 2$, i.e.

$$\frac{1}{1}-\frac{1}{2}+\frac{1}{3}-\frac{1}{4}+\frac{1}{5}-\frac{1}{6}+\frac{1}{7}-\frac{1}{8}+\frac{1}{9}-\frac{1}{10}+\cdots=\ln 2. \qquad (*)$$

This series is called **alternating harmonic series**, which in contrast to the harmonic series, converges to the finite number $\ln 2$.

By property (4-3), with $c=\frac{1}{2}$, we have,

$$\frac{1}{2}-\frac{1}{4}+\frac{1}{6}-\frac{1}{8}+\frac{1}{10}-\frac{1}{12}+\frac{1}{14}-\frac{1}{16}+\frac{1}{18}-\frac{1}{20}+\cdots=\frac{1}{2}\ln 2, \qquad (**)$$

Or equivalently,

$$0+\frac{1}{2}+0-\frac{1}{4}+0+\frac{1}{6}+0-\frac{1}{8}+0+\frac{1}{10}+0-\frac{1}{12}+0+\cdots=\frac{1}{2}\ln 2. \qquad (***)$$

Adding term wise (*) and (***), we have,

$$1+0+\frac{1}{3}-\frac{1}{2}+\frac{1}{5}+0+\frac{1}{7}-\frac{1}{4}+\frac{1}{9}+0+\frac{1}{11}-\frac{1}{6}+\cdots=\frac{3}{2}\ln 2\text{, or}$$

$$1+\frac{1}{3}-\frac{1}{2}+\frac{1}{5}+\frac{1}{7}-\frac{1}{4}+\frac{1}{9}+\frac{1}{11}-\frac{1}{6}+\cdots=\frac{3}{2}\ln 2. \qquad (****)$$

But the series (****) can be obtained from (*), **by rearranging the terms of the latter** (two positive terms followed by one negative), and therefore we see that **the sum of (*) has been altered.**

Note: This Example shows that in general, rearranging the terms of a convergent series, its sum is affected. However, **in a convergent series with positive terms, we can arbitrarily rearrange its terms without affecting its sum**. Let the reader prove it, or see Example 7-6.

Example 4-4.

Find the sum of the series $\sum_{n=1}^{\infty}\frac{3\cdot 2^n-4\cdot 3^n}{5^n}$.

Solution

By property (4-4), we have

$\sum_{n=1}^{\infty} \frac{3 \cdot 2^n - 4 \cdot 3^n}{5^n} = \sum_{n=1}^{\infty} \left\{3 \cdot \left(\frac{2}{5}\right)^n - 4 \cdot \left(\frac{3}{5}\right)^n\right\} = 3 \sum_{n=1}^{\infty} \left(\frac{2}{5}\right)^n - 4 \sum_{n=1}^{\infty} \left(\frac{3}{5}\right)^n =$
$3 \frac{(2/5)}{1-(2/5)} - 4 \frac{(3/5)}{1-(3/5)} = 3 \cdot \frac{2}{3} - 4 \cdot \frac{3}{2} = 2 - 6 = -4.$

Example 4-5.

Investigate the nature of the series,

$$-10^5 - 10^4 - 10^3 - 10^2 - 10 + 1 + \frac{1}{2} + \frac{1}{4} + \frac{1}{6} + \frac{1}{8} + \frac{1}{10} + \frac{1}{12} + \cdots.$$

Solution

By Theorem 4-2, if we delete a finite number of terms from the given series, **we do not alter its nature.** If we delete the first six terms, we obtain a new series,

$$\frac{1}{2} + \frac{1}{4} + \frac{1}{6} + \frac{1}{8} + \frac{1}{10} + \frac{1}{12} + \cdots = \frac{1}{2}\left(1 + \frac{1}{2} + \frac{1}{3} + \frac{1}{4} + \frac{1}{5} + \frac{1}{6} + \cdots\right),$$

which obviously tends to $+\infty$, (Harmonic Series), and therefore the given series diverges to $+\infty$ as well.

PROBLEMS

4-1) Show that removing brackets in **a convergent series with positive terms,** does not alter the sum of the series.

4-2) Show that the series $\sum_{n=1}^{\infty} \left(1 + \frac{1}{n}\right)^n$ diverges to $+\infty$.

4-3) Investigate the nature of the series $\sum_{k=1000}^{\infty} \frac{10^{-5}}{k}$.

(**Answer:** Diverges to $+\infty$).

4-4) Show that the series with general term $u_n = 10^{\left(\frac{3}{n}\right)}$ diverges.

Hint: If $p > 0$ is any fixed positive number, then $\lim \sqrt[n]{p} = 1$.

4-5) Consider the **alternating harmonic series**, (see Example 4-3),rearrange its terms so that each positive term is followed by two negative ones, and show that the sum of the latter series is altered to $\frac{1}{2}$ of the sum of the original.

4-6) Evaluate the sum of the following series,

a) $\sum_{n=3}^{\infty}\frac{-7\cdot 5^n+3\cdot 4^n}{7^n}$ and **b)** $\sum_{n=5}^{\infty}\frac{1+2\cdot 4^n}{5^{2n}}$

4-7) Investigate the nature of the series, having general term

$$u_n=\frac{1}{\sqrt{n+1}-1}-\frac{1}{\sqrt{n+1}+1}.$$

(**Answer:** Diverges to $+\infty$).

4-8) Investigate the nature of the series $\sum_{k=500}^{\infty}\left(1-\frac{1}{k^2}\right)^k$.

4-9) Find the sum of the series $\sum_{k=1}^{\infty}\frac{1}{(2k-1)2k(2k+1)}$.

Hint: Apply partial fractions decomposition, and make use of (*) in Example 4-3.

(**Answer:** $\frac{\ln 4-1}{2}$).

4-10) Consider the series with general term $u_n=\frac{1+2^n}{n!}\cdot(\ln 2)^n$, and show that $\sum_{n=1}^{\infty}u_n=4$.

Hint: If x is any real number,

$$e^x=1+\frac{x}{1!}+\frac{x^2}{2!}+\frac{x^3}{3!}+\cdots=\sum_{n=0}^{\infty}\frac{x^n}{n!},\ \textbf{(0! = 1 by definition).}$$

5. Convergence Criteria for Series with Positive Terms.

In this chapter we consider **infinite series having positive terms only**, and develop some general convergence criteria for such series.

Let us consider the series with **positive terms,**

$\sum_{n=1}^{\infty} \boldsymbol{u_n} = \boldsymbol{u_1} + \boldsymbol{u_2} + \boldsymbol{u_3} + \cdots + \boldsymbol{u_n} + \cdots$, **where all** $\boldsymbol{u_n} > 0$, (5-1)

and the sequence (s_n) of its partial sums,
$\{s_1 = u_1,\ s_2 = u_1 + u_2,\ s_3 = u_1 + u_2 + u_3,\ \cdots\}$.

The sequence (s_n) is increasing, (since $\boldsymbol{s_{n+1}} - \boldsymbol{s_n} = \boldsymbol{u_{n+1}} > 0$). If in addition, (s_n) happens to be bounded above, by some positive constant number M, then according to the well known **Theorem about monotone and bounded sequences,**

$\lim \boldsymbol{s_n} = \boldsymbol{s} \leq \boldsymbol{M} \Leftrightarrow \sum_{n=1}^{\infty} \boldsymbol{u_n} = \boldsymbol{s} \leq \boldsymbol{M}$, (5-2)

meaning that **the infinite series converges to a finite number $s \leq M$.**

If there is no upper bound for (s_n), then $\sum_{n=1}^{\infty} u_n = +\infty$.

In short, **a series with positive terms, either converges to a finite positive number, or diverges to $+\infty$.**

Based on this simple observation, one possible method to investigate the nature of the series in (5-1) is to compare its terms against the terms of **an auxiliary series, whose nature is known.**

Theorem 5-1 (The Comparison Test).

Let $\sum u_n$ and $\sum \phi_n$ be two series with positive terms, such that

$\boldsymbol{0} \leq \boldsymbol{u_n} \leq \boldsymbol{\phi_n}, \forall \boldsymbol{n} \in \mathbb{N}$.

a) If $\sum \phi_n < \infty$, then $\sum u_n < \infty$, as well.

b) If $\sum u_n = +\infty$, then $\sum \phi_n = +\infty$, as well.

Proof: a) Let (s_n) and (w_n) be the sequences of partial sums of $\sum u_n$ and $\sum \phi_n$ respectively. Both (s_n) and (w_n) are increasing sequences.

Since $\sum \phi_n$ converges, by assumption, let $\sum_{n=1}^{\infty} \phi_n = \phi$ **(a finite positive number)**. The term

$s_n = u_1 + u_2 + \cdots + u_n \le \phi_1 + \phi_2 + \cdots + \phi_n < \sum_{n=1}^{\infty} \phi_n = \phi, \quad \forall n$ (*)

meaning that **the increasing sequence (s_n) has the number ϕ as an upper bound**, therefore

$\lim s_n \le \phi \Leftrightarrow \sum_{n=1}^{\infty} u_n \le \phi \Rightarrow \sum_{n=1}^{\infty} u_n$ **converges.** (**)

At the same time, (**) shows that the sum of $\sum_{n=1}^{\infty} u_n$ is less than or equal to the sum $\sum_{n=1}^{\infty} \phi_n = \phi$, (the auxiliary series).

b) If $\sum_{n=1}^{\infty} u_n = +\infty$, then **the sequence (s_n) is unbounded** and therefore (w_n) will be unbounded as well,(from (*)), meaning that $\sum_{n=1}^{\infty} \phi_n = +\infty$.

Note: The comparison test, (and other convergence tests to be derived shortly), applies also in cases where the given condition $0 < u_n \le \phi_n$ **holds from some point on**, for instance $\forall n > 1000$, **and not necessarily from $n = 1$ on.**

For example, if $0 < u_n \le \phi_n \quad \forall n > 1000$, and if $\sum \phi_n$ converges, then $\sum u_n$ converges as well. This is so because **the finite sum $(u_1 + u_2 + \cdots + u_{999} + u_{1000})$ is a fixed number, not affecting the convergence or the divergence of the series $\sum u_n$.**

The **final tail** of the series, $(u_{1001} + u_{1002} + u_{1003} + \cdots + \cdots)$ is important and determines **the nature of the series, (convergence or divergence).**

For this reason, we may sometimes omit the indices in the Σ notation, and write for brevity just $\sum u_n$. The indices, however, should be included, if the sum of the series is to be determined, in which case we have to know how many terms will be included in the summation,(see also Theorem 4-2).

Theorem 5-1 implies next Theorem, known as the limit comparison Test.

Theorem 5-2 (The Limit Comparison Test).

Let $\sum u_n$ and $\sum \phi_n$ be two series with positive terms.

If $\lim\left(\frac{u_n}{\phi_n}\right) = c$, where $c \neq 0$ and $c \neq \infty$, then both series are of the same nature, i.e. either both are convergent or both are divergent to $+\infty$.

Proof: Let $\lim\left(\frac{u_n}{\phi_n}\right) = c > 0$. This means that

$$\forall \varepsilon > 0 \quad \exists N = N(\varepsilon):\ \forall n > N \Rightarrow \left|\frac{u_n}{\phi_n} - c\right| < \varepsilon. \qquad (*)$$

We choose ε very small, so that $c - \varepsilon > 0$. (**)

From (*) and (**) we have,

$$-\varepsilon < \frac{u_n}{\phi_n} - c < \varepsilon \Leftrightarrow 0 < (c-\varepsilon)\phi_n < u_n < (c+\varepsilon)\phi_n, \quad \forall n > N. \ (***)$$

If $\sum \phi_n < \infty$, $(c+\varepsilon)\sum \phi_n < \infty$, and since $0 < u_n < (c+\varepsilon)\phi_n \quad \forall n > N$, Theorem 5-1 implies that $\sum u_n < \infty$, i.e. **the series $\sum u_n$ is convergent.**

If $\sum \phi_n = +\infty$, $(c-\varepsilon)\sum \phi_n = +\infty$, and since $0 < (c-\varepsilon)\phi_n < u_n$, for all $n > N$, again by Theorem 5-1, **the series $\sum u_n$ diverges to $+\infty$.**

Using similar arguments, one may easily show that if $\sum u_n < \infty$ then $\sum \phi_n < \infty$, while if $\sum u_n = +\infty$ then $\sum \phi_n = +\infty$ as well.

Theorems 5-1 and 5-2 are simple, but they suffer a major disadvantage, that is **they require an auxiliary series of known nature**, in order to be applied.

The following Theorems investigate divergence or convergence of a given series, **making use of the general term of the series, only**.

Theorem 5-3 (D' Alembert's test).

Let $\sum u_n$ be a series with positive terms, and let $\lim \frac{u_{n+1}}{u_n} = c$. Then,

a) If $0 \leq c < 1$, the series converges,

b) If $c > 1$, the series diverges to $+\infty$, while

c) If $c = 1$, the Test is inconclusive, that is the series could be either convergent or divergent. (In this case we cannot tell, we have to use another Test).

Proof: Let $\lim\left(\frac{u_{n+1}}{u_n}\right) = c$. Since $u_n > 0$, c will be either zero or positive,(cannot be negative).From the basic definition of limits, we have,

$\forall\varepsilon > 0 \quad \exists N = N(\varepsilon)\colon \quad \forall n > N \Rightarrow \left|\frac{u_{n+1}}{u_n} - c\right| < \varepsilon$, or

$$(c - \varepsilon) < \frac{u_{n+1}}{u_n} < (c + \varepsilon), \qquad \forall n > N. \qquad (*)$$

a) Assuming that $c < 1$, we may choose ε so small that $(\boldsymbol{c} + \boldsymbol{\varepsilon}) < 1$, or if we call $r = c + \varepsilon$,

$$\boldsymbol{r = c + \varepsilon, \quad 0 < r < 1.} \qquad (**)$$

From (*) and (**), we have $\boldsymbol{0 < u_{n+1} < r u_n}, \quad \forall \boldsymbol{n} > N,$

and applying it for $n = N + 1, N + 2, N + 3, \cdots$,we obtain

$0 < u_{N+2} < r u_{N+1},$

$0 < u_{N+3} < r u_{N+2} < r^2 u_{N+1},$

$0 < u_{N+4} < r u_{N+3} < r^3 u_{N+1},$

$\vdots \quad \vdots \quad \vdots \quad \vdots$

Adding term wise the inequalities above, we have,

$$0 < u_{N+1} + u_{N+2} + u_{N+3} + u_{N+4} + \cdots < u_{N+1}(1 + r + r^2 + r^3 + \cdots) =$$
$$= u_{N+1}\frac{1}{1-r}, \qquad (***)$$

since $1 + r + r^2 + r^3 + \cdots = \frac{1}{1-r}$ if $|r| < 1$, (see Theorem 2-1).

Inequality (***) shows that the infinite series $(u_{N+1} + u_{N+2} + u_{N+3} + \cdots)$ converges to a finite limit, less than $\frac{1}{1-r}u_{N+1} > 0$, and therefore the series $(u_1 + u_2 + u_3 + \cdots + u_N) + (u_{N+1} + u_{N+2} + u_{N+3} + \cdots)$ **is also convergent,** and the proof is completed.

b) Let us now assume that $c > 1$. Choosing ε so small, that $\boldsymbol{c - \varepsilon > 1}$, or if we call $\boldsymbol{b = c - \varepsilon}$, $\boldsymbol{b} > 1$, (*) implies

$\frac{u_{n+1}}{u_n} > b > 1 \Longrightarrow u_{n+1} > u_n \quad \forall n > N$, i.e. $u_{N+2} > u_{N+1}$, $u_{N+3} > u_{N+2}$, $u_{N+4} > u_{N+3}$, $\cdots$, i.e. **each term is greater than its predecessor**, and since all the terms of the series are positive, the necessary condition for convergence **($\lim u_n = 0$) is violated**, therefore $\sum u_n = +\infty$.

When D' Alembert's test fails to provide an answer, we may apply the following Theorem, known as **the Cauchy's $n^{\underline{th}}$ root test.**

Theorem 5-4 (Cauchy's $n^{\underline{th}}$ Root Test).

Let $\sum u_n$ be a series with positive terms, and let $\lim \sqrt[n]{u_n} = c$. Then,

a) If $0 \leq c < 1$, the series converges,

b) If $\boldsymbol{c > 1}$, the series diverges to $+\infty$, while

c) If $c = 1$, the Test is inconclusive, that is the series could be either convergent or divergent.(In this case we cannot tell, we have to use another Test).

For a proof see Problem 5-1.

Both D' Alembert's Test and Cauchy's $n^{\underline{th}}$ Root Test fail, if $\lim \frac{u_{n+1}}{u_n} = 1$, or $\lim \sqrt[n]{u_n} = 1$, respectively. In such cases, we may try the so called Raabe's Test, which we state hereunder, without proof.

Theorem 5-5 (Raabe's Test).

Let $\sum u_n$ be a series with positive terms, and let $\lim \left\{ n \left(\frac{u_n}{u_{n+1}} - 1 \right) \right\} = \ell$. Then,

a) If $\ell > 1$, the series converges, while

b) If $\ell < 1$, the series diverges to $+\infty$.

Theorem 5-6 (Generalized Raabe's Test).

Let $\sum u_n$ be a series with positive terms, and let

$$\lim\left\{n\left(\frac{u_n}{u_{n+1}}-1\right)-1\right\}\ln n=\ell.$$ **Then,**

a) If $\ell>1$, the series converges, while

b) If $\ell<1$, the series diverges to $+\infty$.

The following Test, known as the **Cauchy's Integral Test**, is an extremely powerful Test, used often in the investigation of series. In order to follow the proof, the reader is assumed to be familiar with elementary Integral Calculus.

Theorem 5-7 (Cauchy's Integral Test).

Let $\sum u_k$ be a series with positive terms, and let $f(x)$ be a function obtained when k is replaced by x, in the formula for u_k. If $f(x)$ is a decreasing function of x, for $x\geq 1$,then the series $\sum u_k$ and the improper integral $\int_1^{\infty} f(x)\,dx$ are of the same nature, i.e. they both converge or both diverge.

This Theorem is important, **since usually the evaluation if $\int_1^{\infty} f(x)\,dx$ is easier than that of the corresponding series $\sum u_n$.**

If the integral converges, so does the series, if the integral diverges so does the corresponding series.

Proof: Let $f(x)$ be a function satisfying the hypotheses of Theorem 5-7, i.e. **$f(x)$ is positive and decreasing for $x\geq 1$**, and
$\{u_1=f(1),u_2=f(2),\ u_3=f(3),\ \cdots,u_n=f(n),\ \cdots\}$.
The rectangles shown in Fig. 5-1 (a) and Fig.5-1 (b) have areas $u_1,u_2,u_3,\cdots,u_{n-1},u_n$ as indicated in the corresponding Figures. In Fig. 5-1 (a), the total area of the shaded rectangles is less than the area under the curve, from $x=1$ to $x=n$, so

$$u_2+u_3+\cdots+u_n<\int_1^n f(x)dx\Rightarrow$$

$$u_1+u_2+u_3+\cdots+u_n<u_1+\int_1^n f(x)dx. \qquad (*)$$

In Fig. 5-1 (b), the total area of the shaded rectangles is greater than the area under the curve, from $x = 1$ to $x = n$, so

$\int_1^n \boldsymbol{f(x)dx < u_1 + u_2 + \cdots + u_{n-1}} \Rightarrow \int_1^n \boldsymbol{f(x)dx < u_1 + u_2 + \cdots + u_{n-1} + u_n}$, since $u_n > 0$. (**)

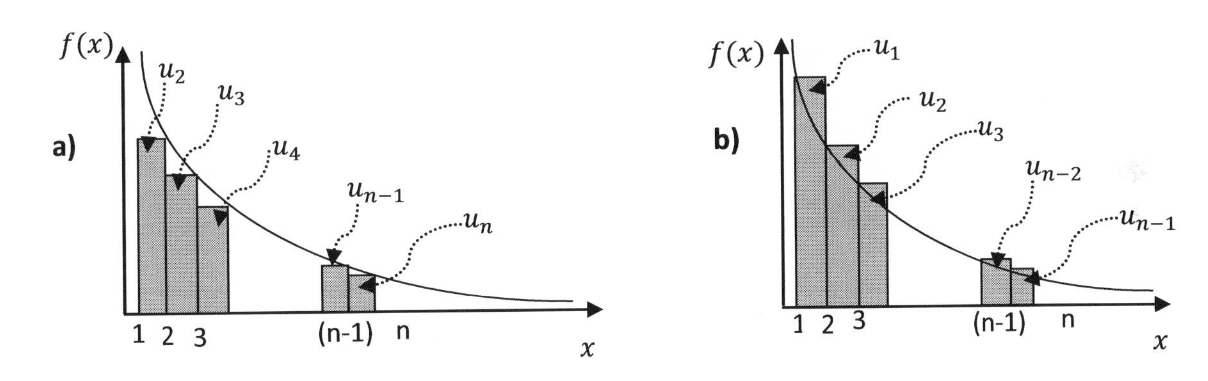

Fig. 5-1: Proof of Cauchy's Integral Test.

From (*) and (**) we have,

$0 < \int_1^n f(x)dx < u_1 + u_2 + u_3 + \cdots + u_n < u_1 + \int_1^n f(x)dx$, or

$0 < \int_1^n f(x)dx < \sum_{k=1}^n u_k < u_1 + \int_1^n f(x)dx$. (***)

Suppose now that we let $n \to \infty$. Then,

i) If the improper integral $\int_1^\infty f(x)dx = \lim_{n\to\infty} \int_1^n f(x)dx$ **converges to a finite number** $\boldsymbol{\ell}$, the corresponding series $\sum_{k=1}^\infty u_k$ converges as well, and furthermore , (from (***),

$\boldsymbol{\ell < \sum_{k=1}^\infty u_k < u_1 + \ell}$, **where** $\boldsymbol{\ell = \int_1^\infty f(x)\, dx}$. (****)

In this case inequality (****) provides also **an estimate** for the sum of the series $\sum_{k=1}^\infty u_k$, which obviously lies between ℓ and $(\ell + u_1)$.

ii) If the improper integral $\int_1^{\infty} f(x)\,dx$ diverges to $+\infty$, again from (***), the corresponding series $\sum_{k=1}^{\infty} u_k$ diverges to $+\infty$ as well, since

$0 < \int_1^{\infty} f(x)dx < \sum_{k=1}^{\infty} u_k$, and this completes the proof of the Theorem.

Next Theorem is also due to Cauchy, and is known as **the Cauchy's Condensation Theorem**.

Theorem 5-8 (Cauchy's Condensation Theorem).

Let $\sum_{n=1}^{\infty} f(n)$ be a series of positive and decreasing terms. Then this series is of the same nature with the series $\sum_{n=1}^{\infty} k^n f(k^n)$, where k is any fixed positive integer > 1.

We note that **the general term of the second series ($k^n f(k^n)$**, is obtained from the general term of the first series (**$f(n)$**, if n is replaced by k^n and then the term $f(k^n)$ is multiplied by k^n, ($k > 1$).

Proof: We want to show that the two series

$\{f(1) + f(2) + f(3) + f(4) + \cdots\}$ and

$\{kf(k) + k^2 f(k^2) + k^3 f(k^3) + k^4 f(k^4) + \cdots\}$

are of the same nature, i.e. the convergence (or divergence) of any one of them, implies the convergence (or the divergence) of the other.

From the first series $\sum f(n)$ let us take the successive terms,

$$\{f(k^m + 1),\ f(k^m + 2),\ f(k^m + 3), \cdots, f(k^{m+1})\}, \qquad (*)$$

where **m is an integer $m \geq 0$.**

The number of terms in (*), is $k^{m+1} - k^m = k^m(k-1)$.

Since $f(x)$ is decreasing, i.e. $\boldsymbol{f(k+1) \leq f(k)}$,

$$k^m(k-1)f(k^{m+1}) < f(k^m + 1) + f(k^m + 2) + \cdots + f(k^{m+1}) < k^m(k-1)f(k^m), \qquad (**)$$

because each term in the summation in (**) is less than $f(k^m)$ and greater than $f(k^{m+1})$. From (**) we have,

$\frac{k-1}{k}k^{m+1}f(k^{m+1}) < f(k^m+1) + f(k^m+2) + \cdots + f(k^{m+1}) <$
$(k-1)k^m f(k^m).$ (***)

Applying (***) for $m = 0,1,2,3,\cdots,n$ yields,

- $\frac{k-1}{k}kf(k) < f(2) + f(3) + \cdots + f(k) < (k-1)f(1),$
- $\frac{k-1}{k}k^2 f(k^2) < f(k+1) + f(k+2) + \cdots + f(k^2) < (k-1)kf(k),$
- $\frac{k-1}{k}k^3 f(k^3) < f(k^2+1) + f(k^2+2) + \cdots f(k^3) < (k-1)k^2 f(k^2),$
- $\vdots \qquad\qquad\qquad \vdots$
- $\frac{k-1}{k}k^{n+1}f(k^{n+1}) < f(k^n+1) + f(k^n+2) + \cdots f(k^{n+1}) <$
  $(k-1)k^n f(k^n).$

Adding the above $(n+1)$ inequalities term wise, we obtain,

$\frac{k-1}{k}\{kf(k) + k^2 f(k^2) + \cdots + k^{n+1}f(k^{n+1})\} < -f(1) + \sum_{\ell=1}^{k^{n+1}} f(\ell) <$
$(k-1)f(1) + (k-1)\{kf(k) + k^2 f(k^2) + \cdots + k^n f(k^n)\}.$ (****)

If we call,

$\boldsymbol{S_n = \sum_{\ell=1}^{k^{n+1}} f(\ell)}$ and $\boldsymbol{T_n = \sum_{\ell=1}^{k^n} k^\ell f(k^\ell)}$, inequality (****) yields,

$\boldsymbol{\frac{k-1}{k}T_n < S_n - f(1) < (k-1)\{T_n + f(1)\}}.$ (*****)

From (*****) we conclude that:

1) If the second series converges ($\lim T_n = \sum_{\ell=1}^{\infty} k^\ell f(k^\ell) < \infty$) then the original series converges as well ($\lim S_n = \sum_{\ell=1}^{\infty} f(\ell) < \infty$), and conversely.

2) If the second series diverges ($\lim T_n = \sum_{\ell=1}^{\infty} k^\ell f(k^\ell) = +\infty$) then the original series also diverges ($\lim S_n = \sum_{\ell=1}^{\infty} f(\ell) = +\infty$), and conversely, and this completes the proof of the Theorem.

Review of the Convergence Tests for Series with Positive Terms:

In our so far analysis, we have developed some powerful **Convergence Tests for series $\sum u_n$ with positive terms, ($u_n > 0$).**

Given a series $\sum u_n$ $(u_n > 0)$, **one of the main problems is to investigate the nature of the series, i.e. to determine whether the series converges to a finite number or if it diverges to** $+\infty$.

The approach towards this direction, in general terms, is indicated below:

- If $\lim u_n \neq 0$, then the series diverges.
- If $\lim u_n = 0$, the series may converge or it may diverge.
- To figure out the nature of the series, we may start applying some **simple Convergence Tests**, i.e. comparison with an auxiliary series of known nature, or apply the D' Alembert's Test, or the Cauchy's Test, etc.
- **All the Convergence Tests, give answer to the main question, regarding the convergence or the divergence of the series**. Once established that a series converges, **the next question which naturally arises, is to evaluate the sum of the series**. This by itself is a difficult problem. In some simple cases, (geometric progression with ratio $|x| < 1$, (see Theorem 2-1), or telescoping series (see Chapter 3)), the sum of the series can be obtained in **closed form**, but for most of the cases, this is not possible.
- On the other hand, series appear quite often, and in a very natural way in many applied problems, in Engineering, Physics, Geometry etc, and finding the sum of such series is a problem of great practical importance. In such cases, **digital computes are used, in order to find the sum of a series, to any desired degree of accuracy, sufficient for all practical applications and problems.**

Example 5-1.

Determine whether the following series converge or diverge.

a) $\sum \frac{n!}{n^n}$ **b)** $\sum \frac{1}{n^n \ln n}$ **c)** $\sum \frac{1}{n}$

Solution

a) Let us investigate this series making use of the **D' Alembert's Test**. The general term is $u_n = \frac{n!}{n^n}$ from which,

$\frac{u_{n+1}}{u_n} = \frac{(n+1)! \div (n+1)^{n+1}}{n! \div n^n} = \frac{\{(n+1)!\}n^n}{n!(n+1)^{n+1}} = \frac{n^n}{(n+1)^n} = \frac{1}{\left(1+\frac{1}{n}\right)^n}$, and

$\lim \frac{u_{n+1}}{u_n} = \lim \frac{1}{\left(1+\frac{1}{n}\right)^n} = \frac{1}{e} < 1$, therefore the given series converges.

Note: Recall that $\lim \left(1 + \frac{1}{n}\right)^n = e = 2.7182 \cdots > 1$, **(Euler's Number).**

b) Let us again apply the **D' Alembert's Test**. In this case $u_n = \frac{1}{n^n \ln n}$ and

$\frac{u_{n+1}}{u_n} = \frac{1}{n+1} \cdot \frac{1}{\left(1+\frac{1}{n}\right)^n} \cdot \frac{\ln n}{\ln(n+1)}$, from which $\lim \frac{u_{n+1}}{u_n} = 0 \cdot \frac{1}{e} \cdot 1 = 0 < 1$,

therefore the series converges.

c) The series $\sum \frac{1}{n}$ is **the Harmonic Series** which as we already know diverges to $+\infty$, (see Theorem 4-6). However in this problem we would like to investigate the nature of this series with the aid of Theorems and Tests developed in this chapter. In our case $u_n = \frac{1}{n}$.

Let us start with the **D' Alembert's Test**, i.e. $\lim \frac{u_{n+1}}{u_n} = \lim \frac{n}{n+1} = 1$, **(inconclusive)**.

Let us next try the **Cauchy's n^{th} Root Test**, i.e.

$\lim \sqrt[n]{u_n} = \lim \sqrt[n]{\frac{1}{n}} = \lim \frac{1}{\sqrt[n]{n}} = \frac{1}{1} = 1$,

since as we know $\lim \sqrt[n]{n} = 1$, and again we can draw no conclusion.

If we try the **Raabe's Test**, (Theorem 5-5), we have,

$\lim \left\{n\left(\frac{u_n}{u_{n+1}} - 1\right)\right\} = \lim \left\{n\left(\frac{n+1}{n} - 1\right)\right\} = \lim \left\{n \cdot \frac{1}{n}\right\} = 1$,

and still no conclusion can be drawn. Let us now try the **Raabe's generalized Test** (Theorem 5-6). In this case,

$\left\{n\left(\frac{u_n}{u_{n+1}} - 1\right) - 1\right\} \ln n = \left\{n\left(\frac{n+1}{n} - 1\right) - 1\right\} \ln n = (n + 1 - n - 1) \ln n = 0$,

therefore the $\lim \left\{n\left(\frac{u_n}{u_{n+1}} - 1\right) - 1\right\} \ln n = 0 < 1$,

and by virtue of Theorem 5-6, the Harmonic Series diverges to $+\infty$.

We note that such a simple series resists all elementary convergence criteria, and its nature is revealed with the application of the rather complicated generalized Raabe's Test.

Note: We can show that the harmonic series diverges, very easily, making use of the **Cauchy's Integral Test**, (Theorem 5-7). The corresponding function $f(x) = \frac{1}{x}$ is positive and decreasing for $x \geq 1$, **the improper integral** $\int_1^\infty \frac{dx}{x} = +\infty$, **(diverges)**, and therefore **the Harmonic Series diverges to** $+\infty$ **as well**.

Example 5-2.

Investigate the nature of the series $\sum_{n=0}^{\infty} \frac{x^n}{n!}$ $(0! = 1$ by definition), where x is any **fixed** real number, ($x \in \mathbb{R}$).

Solution

The general term u_n of the series

$\sum_{n=0}^{\infty} \frac{x^n}{n!} = 1 + \frac{x}{1!} + \frac{x^2}{2!} + \frac{x^3}{3!} + \frac{x^4}{4!} + \cdots$, is $u_n = \frac{x^n}{n!}$.

Application of the D' Alembert's Test yields,

$$\frac{u_{n+1}}{u_n} = \frac{x^{n+1} \div (n+1)!}{x^n \div n!} = \frac{x}{n+1} \quad \text{and} \quad \lim \frac{u_{n+1}}{u_n} = \lim \frac{x}{n+1} = x \cdot \lim \frac{1}{n+1} = x \cdot 0 = 0,$$

for every real value of x.

This means that **the given series converges for all real values of x.**

Note: The given series, which as it was proved **converges for all real values of x**, defines the so called **exponential function e^x**, i.e.

$$e^x = \sum_{n=0}^{\infty} \frac{x^n}{n!} = 1 + \frac{x}{1!} + \frac{x^2}{2!} + \frac{x^3}{3!} + \frac{x^4}{4!} + \cdots, \qquad -\infty < x < +\infty.$$

In particular,

For $x = 1$, $e = 1 + \frac{1}{1!} + \frac{1}{2!} + \frac{1}{3!} + \frac{1}{4!} + \cdots = 2.7182\cdots$,

is the Euler's number, while

For $x = -1$, $e^{-1} = 1 - \frac{1}{1!} + \frac{1}{2!} - \frac{1}{3!} + \frac{1}{4!} - \cdots = 0.3678\cdots$.

The number e is also the limit of the sequence $\left(1+\frac{1}{n}\right)^n$ i.e. $\lim\left(1+\frac{1}{n}\right)^n = e$ while $\lim\left(1-\frac{1}{n}\right)^n = e^{-1}$, (see Example 5-12).

The number e (like the number π), is a transcendental number (proved by Charles Hermite in 1873) meaning that there exists no polynomial with integer coefficients, admitting e as a root.

Example 5-3. (The p -series).

Let us consider the series

$$\sum_{n=1}^{\infty}\frac{1}{n^p} = \frac{1}{1^p}+\frac{1}{2^p}+\frac{1}{3^p}+\frac{1}{4^p}+\frac{1}{5^p}+\cdots, \text{ where } p>0. \qquad (*)$$

The series (*) is known as the **p - series.**

For example,

For $p=1$ we obtain the Harmonic Series $\frac{1}{1}+\frac{1}{2}+\frac{1}{3}+\frac{1}{4}+\frac{1}{5}+\cdots$,

For $p=2$ we obtain the series $\frac{1}{1^2}+\frac{1}{2^2}+\frac{1}{3^2}+\frac{1}{4^2}+\frac{1}{5^2}+\cdots$,

For $p=\frac{1}{2}$ we obtain the series $\frac{1}{\sqrt{1}}+\frac{1}{\sqrt{2}}+\frac{1}{\sqrt{3}}+\frac{1}{\sqrt{4}}+\frac{1}{\sqrt{5}}+\cdots$, etc.

In order to investigate the nature of the p-series, we shall make use of the Cauchy's Integral Test, (Theorem 5-7). The function $f(x)$ in this case is $f(x)=x^{-p},\ p>0$, which is positive and decreasing for $x\geq 1$. Let us now consider the improper integral,

$$\int_1^{\infty} x^{-p}dx = \lim_{M\to\infty}\int_1^{M} x^{-p}\,dx = \lim_{M\to\infty}\frac{x^{-p+1}}{-p+1} =$$

$$= \frac{1}{1-p}\lim_{M\to\infty}\{M^{-p+1}-1\}, \qquad (**)$$

where **we have assumed that $p\neq 1$.**

From (**) we have:

1) If $1-p>0 \Leftrightarrow 0<p<1$, then $\lim_{M\to\infty} M^{1-p} = +\infty$,
i.e. the $\int_1^{\infty} f(x)\,dx = +\infty$, and from Theorem 5-7, the series $\sum_{n=1}^{\infty}\frac{1}{n^p}$ diverges for $0<p<1$.

2) If $1-p<0 \Leftrightarrow p>1$, the $\lim_{M\to\infty} M^{1-p}=0$,
i.e. the $\int_1^{\infty} x^{-p} = \frac{1}{p-1} < +\infty$, and again from Theorem 5-7, the series $\sum_{n=1}^{\infty} \frac{1}{n^p}$ converges for $p>1$.

Finally, for $p=1$, **the p −series becomes the Harmonic Series**, which as we have already proved diverges to $+\infty$.

In summary,

The p −series $\sum_{n=1}^{\infty} \frac{1}{n^p}$ with $p>0$,

- **Converges if $p>1$, and**
- **Diverges if $0<p\le 1$.**

This result is very important, since the p −series can now be used as **an auxiliary series of known nature**, in the investigation of more complicated series.

Furthermore, from (****) in Theorem 5-7, for $p>1$,

$\frac{1}{p-1} < \sum_{n=1}^{\infty} \frac{1}{n^p} < 1+\frac{1}{p-1} = \frac{p}{p-1}$, **(since** $\int_1^{\infty} x^{-p} dx = \frac{1}{p-1}$**)**.

Note: As we have just proved, **the series $\sum_{n=1}^{\infty} \frac{1}{n^p}$ converges if $p>1$**. This means that for each positive $p>1$, we can associate a number, **which is the sum of the corresponding p −series**, and we denote this number as $\zeta(p)$, i.e.

$$\zeta(p) = \sum_{n=1}^{\infty} \frac{1}{n^p} \qquad p>1.$$

The function thus defined, for all $p>1$, is known as **the Riemann's zeta function**, after the great German mathematician **Bernard Riemann (1826-1866),** although this function was firstly introduced by **Leonard Euler,(one of the greatest mathematicians of all times), in 1737.**

It comes as a surprise (and is quite unexpected) that **the values of $\zeta(p)$, at even values of p ($p=2,4,6,\cdots$), relate to the number π**. For example, it can be shown that

$$\sum_{n=1}^{\infty}\frac{1}{n^2}=\frac{\pi^2}{6},\ \sum_{n=1}^{\infty}\frac{1}{n^4}=\frac{\pi^4}{90},\ \sum_{n=1}^{\infty}\frac{1}{n^6}=\frac{\pi^6}{945},\cdots$$

In particular the first result ($\sum_{n=1}^{\infty}\frac{1}{n^2}=\frac{\pi^2}{6}$) is the famous **'Basel's Problem', solved for the first time by Euler, in 1730**, by a method of great ingenuity and insight, which earned him worldwide reputation and fame.

For a detailed account on the Basel's Problem, see Example 11-5.

The values of $\zeta(p)$ at odd values of p $(3,5,7,\cdots)$ are not known in closed form. This is still an open problem.

Example 5-4.

Determine whether the following series are convergent or divergent,

a) $\sum\frac{1}{\sqrt{n^3}}$ **b)** $\sum\frac{n+3}{n^4+n^2+5}$ **c)** $\sum_{n=2}^{\infty}\frac{1}{n(\ln n)^a}$ $a>0.$

Solution

a) $\sum\frac{1}{\sqrt{n^3}}=\sum\frac{1}{n^{(3/2)}}.$

This is a p −series with $p=\frac{3}{2}>1$, and is therefore convergent.

b) Let $u_n=\frac{n+3}{n^4+n^2+5}$ be the general term of the given series, and let us consider **an auxiliary series**, with general term $\phi_n=\frac{1}{n^3}$. The series $\sum\phi_n$ converges, since it is a p −series with $p=3>1$.

The $\lim\left(\frac{u_n}{\phi_n}\right)=\lim\frac{n^4+3n^3}{n^4+n^2+5}=1$, and according to Theorem 5-2, the series $\sum u_n$ is convergent

c) To investigate the nature of this series, we shall apply **the Cauchy's Integral Test**. The function $f(x)=\frac{1}{x(\ln x)^a}$ where $a>0$, is positive and decreasing for $x\geq 2$. For $\boldsymbol{a}\neq\mathbf{1}$, we have,

$$\int_2^{\infty}\frac{dx}{x(\ln x)^a}=\int_2^{\infty}\frac{d(\ln x)}{(\ln x)^a}=\int_2^{\infty}(\ln x)^{-a}d(\ln x)=\frac{1}{1-a}(\ln x)^{1-a}\Big|_2^{\infty}\quad(*)$$

1) If $(1-a)>0$, i.e. if $\mathbf{0<a<1}$, $\frac{1}{1-a}(\ln x)^{1-a}\Big|_2^{\infty}=+\infty$,

and from (*), $\int_2^{\infty}\frac{dx}{x(\ln x)^a}=+\infty$, meaning that the corresponding series diverges to $+\infty$.

2) If $(1-a)<0$, i.e. if $\mathbf{a>1}$, $\frac{1}{1-a}(\ln x)^{1-a}\Big|_2^{\infty}=-\frac{1}{1-a}(\ln 2)^{1-a}$,

and from (*), $\int_2^{\infty}\frac{dx}{x(\ln x)^a}=\frac{1}{a-1}(\ln 2)^{1-a}<+\infty$, and therefore the corresponding series converges as well.

Finally, if $\mathbf{a=1}$, the integral becomes

$$\int_2^{\infty}\frac{dx}{x\ln x}=\int_2^{\infty}\frac{d(\ln x)}{\ln x}=\ln(\ln x)\big|_2^{\infty}=+\infty.$$

In summary,

the series $\sum_{n=2}^{\infty}\frac{1}{n(\ln n)^a}$ $\quad a>0$, **diverges if** $\mathbf{0<a\leq 1}$ **and converges if** $\mathbf{a>1}$.

Example 5-5.

Investigate the following series

a) $\sum\left(\sqrt{n^4+1}-n^2\right)$ **b)** $\sum\left(\sqrt{n^3+1}-\sqrt{n^3}\right)$.

Solution

a) Let $u_n=\sqrt{n^4+1}-n^2=\frac{(n^4+1)-n^4}{\sqrt{n^4+1}+n^2}=\frac{1}{\sqrt{n^4+1}+n^2}$.

If we consider **the auxiliary series** $\sum\phi_n$ **with** $\phi_n=\frac{1}{n^2}$ we have,

$\lim\left(\frac{u_n}{\phi_n}\right)=\lim\frac{n^2}{\sqrt{n^4+1}+n^2}=\lim\frac{1}{\sqrt{1+\frac{1}{n^4}}+1}=\frac{1}{2}$ and since $\sum\phi_n<\infty$

(**p –series with $p=2>1$**), the $\sum u_n<\infty$, by virtue of Theorem 5-2.

b) In this case,

$$u_n=\sqrt{n^3+1}-\sqrt{n^3}=\frac{(n^3+1)-n^3}{\sqrt{n^3+1}+\sqrt{n^3}}=\frac{1}{\sqrt{n^3+1}+\sqrt{n^3}}.$$

Considering the convergent auxiliary series $\sum \phi_n$ with $\phi_n = \frac{1}{n^{(3/2)}}$

($\boldsymbol{p}$ **–series with** $\boldsymbol{p} = \frac{3}{2} > 1$), and noting that $\lim \left(\frac{u_n}{\phi_n}\right) = \frac{\sqrt{n^3}}{\sqrt{n^3+1}+\sqrt{n^3}} = \frac{1}{2}$,

we conclude that $\sum u_n$ converges, by Theorem 5-2.

Example 5-6. (The Euler-Mascheroni constant $\gamma \cong 0.57721$).

Show that the quantity

$$\boldsymbol{u_n = \sum_{k=1}^{n} \frac{1}{k} - \ln n = 1 + \frac{1}{2} + \frac{1}{3} + \frac{1}{4} + \frac{1}{5} + \cdots + \frac{1}{n} - \ln n}$$

tends to a finite limit as $n \to \infty$.

Solution

Let us consider the graph of the function $y = \frac{1}{x}$ for $x \geq 1$, as shown in Fig. 5-2.

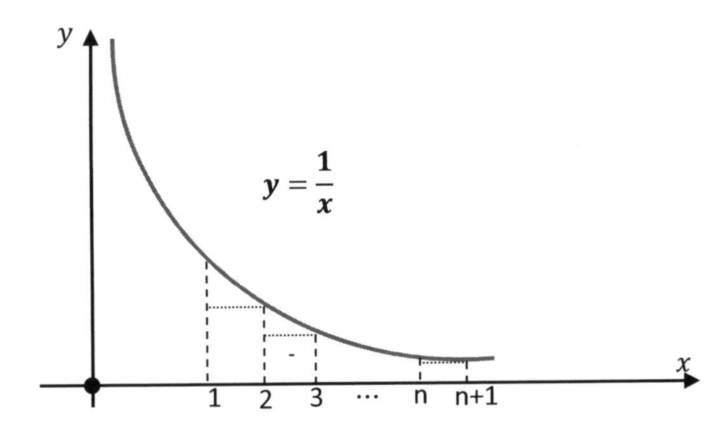

Fig.5-2: Graph of the function $y = \frac{1}{x}$ $x > 0$.

From this Figure, one easily obtains (see also Proof of Theorem 5-7),

$$\frac{1}{2}\cdot 1 + \frac{1}{3}\cdot 1 + \cdots + \frac{1}{n}\cdot 1 + \frac{1}{n+1}\cdot 1 < \int_1^{n+1}\frac{dx}{x} < \frac{1}{1}\cdot 1 + \frac{1}{2}\cdot 1 + \cdots + \frac{1}{n-1}\cdot 1 + \frac{1}{n}\cdot 1$$

or

$\frac{1}{2}+\frac{1}{3}+\cdots+\frac{1}{n}+\frac{1}{n+1}<\ln(n+1)<1+\frac{1}{2}+\frac{1}{3}+\cdots+\frac{1}{n-1}+\frac{1}{n},$

from which one obtains,

$$0<1+\frac{1}{2}+\frac{1}{3}+\cdots+\frac{1}{n-1}+\frac{1}{n}-\ln(n+1)<1-\frac{1}{n+1}. \qquad (*)$$

If we define $\boldsymbol{s_n}=\mathbf{1}+\frac{\mathbf{1}}{\mathbf{2}}+\frac{\mathbf{1}}{\mathbf{3}}+\cdots+\frac{\mathbf{1}}{\boldsymbol{n}-\mathbf{1}}+\frac{\mathbf{1}}{\boldsymbol{n}}-\mathbf{ln}(\boldsymbol{n}+\mathbf{1})$,

inequality (*) implies ,

$$0<s_n<1-\frac{1}{n+1}<1, \qquad (**)$$

meaning that **the sequence (s_n) is bounded between 0 and 1**, for all values of n. At the same time **the sequence (s_n) is increasing**. To show this, we consider the difference $s_{n+1}-s_n=\frac{1}{n+1}+\ln(n+1)-\ln(n+2)$. (***)

Application of the **Mean Value Theorem** to the function $y=\ln x$, between the points $(n+1)$ and $(n+2)$, yields,

$$\ln(n+1)-\ln(n+2)=(-1)\frac{1}{\xi}, \text{ where } (n+1)<\xi<(n+2),$$

and (***) becomes ,

$s_{n+1}-s_n=\frac{1}{n+1}-\frac{1}{\xi}=\frac{\xi-(n+1)}{\xi(n+1)}>0$, meaning that **the sequence (s_n) is increasing.**

Therefore **the sequence (s_n) being increasing and bounded above by the number 1**, (as indicated in (**)), **tends to a limit ≤ 1, according to the fundamental Theorem about monotone and bounded sequences.**

We have thus far shown that the $\lim s_n$ exists and is smaller than or equal to 1. The $n^{\underline{th}}$ term of (s_n) is

$$s_n=1+\frac{1}{2}+\frac{1}{3}+\cdots+\frac{1}{n}-\ln(n+1)=1+\frac{1}{2}+\frac{1}{3}+\cdots+\frac{1}{n}-\ln\left(n\left(1+\frac{1}{n}\right)\right)=$$
$$1+\frac{1}{2}+\frac{1}{3}+\cdots+\frac{1}{n}-\ln n-\ln\left(1+\frac{1}{n}\right)\Rightarrow 1+\frac{1}{2}+\frac{1}{3}+\cdots\frac{1}{n}-\ln n=$$
$$s_n+\ln\left(1+\frac{1}{n}\right)\Rightarrow \boldsymbol{u_n}=\boldsymbol{s_n}+\mathbf{ln}\left(\mathbf{1}+\frac{\mathbf{1}}{\boldsymbol{n}}\right). \qquad (****)$$

Since the $\mathbf{lim}\,\boldsymbol{s_n}$ **exists** and the $\lim\left(\ln\left(1+\frac{1}{n}\right)\right)=\ln(1+0)=\ln 1=0$, taking the limit of both sides in (****) as $n\to\infty$, implies that

$\lim \boldsymbol{u_n} = \lim \left\{1 + \frac{1}{2} + \frac{1}{3} + \cdots + \frac{1}{n} - \ln n\right\} = \lim s_n \leq 1$, and this completes the proof.

Note: In a completely similar manner, one may show that the sequence

$\boldsymbol{u_n} = \mathbf{1} + \frac{1}{2} + \frac{1}{3} + \cdots \frac{1}{n} - \ln n$ **is decreasing** ,i.e. $u_{n+1} < u_n \quad \forall n$, (see Problem 5-4). Also it is easily shown that $\boldsymbol{u_n} > \boldsymbol{s_n} \quad \forall \boldsymbol{n}$ and that the $\lim(\boldsymbol{u_n} - \boldsymbol{s_n}) = \mathbf{0}$, (see Problem 5-5). According to **the principle of nested intervals**, both sequences converge to **a common limit**, which is called **the Euler-Mascheroni constant and denoted by the Greek letter** γ, i.e.

$$\gamma = \lim \left(1 + \frac{1}{2} + \frac{1}{3} + \cdots + \frac{1}{n} - \ln n\right) =$$

$$= \lim \left(1 + \frac{1}{2} + \frac{1}{3} + \cdots + \frac{1}{n} - \ln(n+1)\right) \cong 0.57721\cdots$$

The approach shown in Figure 5-3 may be used, if one wishes to find approximate values for the constant γ.

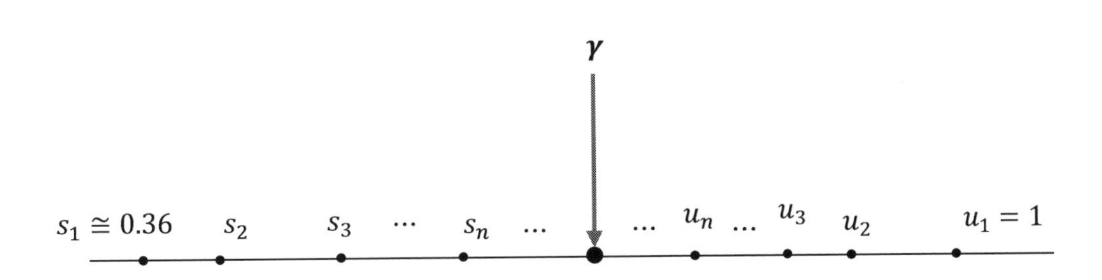

Fig.5-3: Obtaining approximate values for γ.

For example,

- For $n = 3$, $\quad s_3 < \gamma < u_3 \Rightarrow 0.446 < \gamma < 0.734$,
- For $n = 5$, $\quad s_5 < \gamma < u_5 \Rightarrow 0.491 < \gamma < 0.673$,
- For $n = 10$, $s_{10} < \gamma < u_{10} \Rightarrow 0.531 < \gamma < 0.626$.

Example 5-7.

Determine whether the series $\sum_{n=1}^{\infty} \frac{x^n}{n^{n+2}}$ is convergent or divergent, if x is any fixed positive number.

Solution

The general term of the given series is $\boldsymbol{u_n} = \frac{\boldsymbol{x^n}}{\boldsymbol{n^{n+2}}}$. We shall make use of the **Cauchy's $n^{\underline{th}}$ Root Test**, (Theorem 5-4) in order to investigate the nature of the series. We have,

$$\sqrt[n]{u_n} = \sqrt[n]{\frac{x^n}{n^{n+2}}} = \frac{x}{n}\left(\frac{1}{\sqrt[n]{n}}\right)^2, \qquad (*)$$

and since $\lim \sqrt[n]{n} = 1$, (*) implies that

$$\lim \sqrt[n]{u_n} = x \cdot \lim\left(\frac{1}{n}\right) \cdot \left(\frac{1}{\lim \sqrt[n]{n}}\right)^2 = x \cdot 0 \cdot 1 = 0 < 1,$$

meaning that **the given series converges for all positive values of x.**

Note: For a solution using D' Alembert's Theorem, see Problem 5-6.

Example 5-8.

Investigate the convergence of the series $\sum \frac{n^k}{n!}$, (is any **fixed** real number).

Solution

Making use of the **D' Alembert's Test** (Theorem 5-3), one easily finds that

$$\lim \frac{u_{n+1}}{u_n} = \lim\left(1 + \frac{1}{n}\right)^k \cdot \lim\left(\frac{1}{n+1}\right) = 1 \cdot 0 = 0 < 1,$$

Therefore the series converges for any real value of k.

Example 5-9.

Test for convergence or divergence,

a) $\sum_{n=2}^{\infty} \frac{1}{2^n} \tan\left(\frac{\pi}{2^n}\right)$ **b)** $\sum_{n=1}^{\infty} \left\{\sin\left(x + \frac{x}{n}\right)\right\}^n$, $0 < x < \frac{\pi}{2}$.

Solution

a) The general term of the series is $\boldsymbol{u_n = \frac{1}{2^n}\tan\left(\frac{\pi}{2^n}\right)}$.

Let us apply the **D' Alembert's Test**.

$$\frac{u_{n+1}}{u_n} = \frac{1}{2}\cdot\frac{\tan\left(\frac{\pi}{2^{n+1}}\right)}{\tan\left(\frac{\pi}{2^n}\right)} = \frac{1}{2}\cdot\frac{\tan\left(\frac{\pi}{2^{n+1}}\right)\div\left(\frac{\pi}{2^{n+1}}\right)}{\tan\left(\frac{\pi}{2^n}\right)\div\left(\frac{\pi}{2^n}\right)}\cdot\frac{1}{2}. \qquad (*)$$

As $n \to \infty$, $\left(\frac{\pi}{2^n}\right) \to 0$ and $\left(\frac{\pi}{2^{n+1}}\right) \to 0$, therefore

$\lim\left\{\tan\left(\frac{\pi}{2^{n+1}}\right) \div \left(\frac{\pi}{2^{n+1}}\right)\right\} = 1$, and $\lim\left\{\tan\left(\frac{\pi}{2^n}\right) \div \left(\frac{\pi}{2^n}\right)\right\} = 1$,

(since $\mathbf{lim}_{y\to 0}\frac{\tan y}{y} = \mathbf{1}$, see Note in Example 3-7),

and from (*), $\lim\frac{u_{n+1}}{u_n} = \frac{1}{4}\cdot 1 = \frac{1}{4} < 1$, meaning that **the given series converges.**

For the **exact evaluation of its sum**, see Problem 5-7.

b) The general term of the series is $\boldsymbol{a_n = \left\{\sin\left(x + \frac{x}{n}\right)\right\}^n}$, where $0 < x < \frac{\pi}{2}$.

We first note that $0 < x + \frac{x}{n} < \frac{\pi}{2}\left(1 + \frac{1}{n}\right) < \pi$, therefore $\sin\left(x + \frac{x}{n}\right) > 0$,

and subsequently $a_n = \left\{\sin\left(x + \frac{x}{n}\right)\right\}^n > 0$.

If we apply the **Cauchy's $\boldsymbol{n^{\underline{th}}}$ Root Test**, we get,

$\sqrt[n]{a_n} = \sin\left(x + \frac{x}{n}\right)$ and the

$\lim\sqrt[n]{a_n} = \lim\left\{\sin\left(x + \frac{x}{n}\right)\right\} = \sin(x + 0) = \sin x < 1$,

therefore **the given series converges**.

Example 5-10.

Making use of the **Cauchy's Condensation Test**, (Theorem 5-8), test for convergence the $\boldsymbol{p}$ **–series** $\sum\frac{\mathbf{1}}{\boldsymbol{n^p}}$, $\boldsymbol{p} > 0$.

Solution

In Example 5-3 we investigated the $\boldsymbol{p}$ **−series** using the Cauchy's Integral Test. In this Example we shall investigate the same series, using the Cauchy's Condensation Test. The p −series is $\sum \frac{1}{n^p}$ where $p > 0$.

In this case $\boldsymbol{f(n)} = \frac{1}{n^p}$ and according to Theorem 5-8, **the** $\boldsymbol{p}$ **−series will be of the same nature with the series** $\sum \boldsymbol{k^n f(k^n)}$**, where** $\boldsymbol{k} > 1$. We have,

$$k^n f(k^n) = k^n \frac{1}{(k^n)^p} = (k^n)^{1-p}, \; k > 1.$$

The corresponding series,

$$\sum_{n=1}^{\infty} k^n f(k^n) = k^{1-p} + (k^{1-p})^2 + (k^{1-p})^3 + \cdots \qquad (*)$$

is a geometric progression with ratio $\boldsymbol{k^{1-p}}$, and converges if $k^{1-p} < 1$, i.e. if $\boldsymbol{p} > 1$.

So for $\boldsymbol{p} > 1$, the series (*) converges and therefore the series $\sum_{n=1}^{\infty} \frac{1}{n^p}$ converges as well.

If $\mathbf{0} < p \leq 1$, the series (*) diverges to $+\infty$, and therefore so does the series $\sum_{n=1}^{\infty} \frac{1}{n^p}$.

Example 5-11.

Investigate for convergence the series with general term

$$\boldsymbol{u_n} = \boldsymbol{a}^{1+\frac{1}{2}+\frac{1}{3}+\cdots+\frac{1}{n}}, \qquad \boldsymbol{a} > 0.$$

Solution

Application of the simple D' Alembert's Test, does not reveal the nature of the series, since $\lim \frac{u_{n+1}}{u_n} = \lim a^{\frac{1}{n+1}} = \sqrt[n+1]{a} = 1$, (**inconclusive**).

Let us now try the Raabe's Test (Theorem 5-5), by considering the quantity

$$Q = n\left(\frac{u_n}{u_{n+1}} - 1\right) = n\left(\frac{1}{a^{\frac{1}{n+1}}} - 1\right),$$ or if we set $\boldsymbol{x} = \frac{\mathbf{1}}{\boldsymbol{n}+\mathbf{1}}$,

$$\boldsymbol{Q = \frac{1-x}{x}\left(\frac{1}{a^x} - 1\right) = (1-x)\frac{\left(\frac{1}{a}\right)^x - 1}{x}}. \qquad (*)$$

As $n \to \infty,\ x \to 0$ and the term $\frac{\left(\frac{1}{a}\right)^x - 1}{x} \to \frac{1-1}{0} = \frac{0}{0}$,

and making use of **the De L' Hopital's rule**

$$\lim_{x\to 0} \frac{\left(\frac{1}{a}\right)^x - 1}{x} = \lim_{x\to 0} \frac{\left(\frac{1}{a}\right)^x \ln\left(\frac{1}{a}\right)}{1} = \ln\left(\frac{1}{a}\right),$$

therefore (*) implies,

$$\lim_{n\to\infty} Q = \lim_{x\to 0} Q = 1 \cdot \ln\left(\frac{1}{a}\right) = \ln\left(\frac{1}{a}\right). \qquad (**)$$

According to **the Raabe's Test,**

If $\lim_{n\to\infty} Q = \ln\left(\frac{1}{a}\right) > 1$, the series will be convergent, i.e.

if $\ln\left(\frac{1}{a}\right) > \ln e \Leftrightarrow \frac{1}{a} > e \Leftrightarrow a < \frac{1}{e}$ **the series converges,**

while if $a > \frac{1}{e}$ **($\ln\left(\frac{1}{a}\right) < 1$) the series diverges**. In summary,

$\sum_{n=1}^{\infty} a^{1+\frac{1}{2}+\frac{1}{3}+\cdots+\frac{1}{n}}$ **converges if $0 < a < \frac{1}{e}$ and diverges if $a > \frac{1}{e}$.**

At $a = \frac{1}{e}$, the $\lim_{n\to\infty} Q = \ln e = 1$, and the Raabe's Test is **inconclusive**. We have to examine this case separately. From Fig.5-2, in Example 5-6, we have,

$$\frac{1}{2}\cdot 1 + \frac{1}{3}\cdot 1 + \cdots + \frac{1}{n}\cdot 1 < \int_1^n \frac{dx}{x} \Longrightarrow \frac{1}{2} + \frac{1}{3} + \cdots + \frac{1}{n} < \ln n, \text{ or}$$

$$-\left(1 + \frac{1}{2} + \frac{1}{3} + \cdots + \frac{1}{n}\right) > -(1 + \ln n). \qquad (***)$$

The given series at $a = \frac{1}{e} = e^{-1}$ becomes $\sum u_n$ where $u_n = e^{-\left(1+\frac{1}{2}+\frac{1}{3}+\cdots+\frac{1}{n}\right)}$, and taking (***) into consideration, we have,

$$u_n > e^{-(1+\ln n)} = e^{-1} \cdot e^{-\ln n} = e^{-1} \cdot e^{\ln\left(\frac{1}{n}\right)} = e^{-1} \cdot \left(\frac{1}{n}\right),$$

and since the harmonic series $\sum \frac{1}{n}$ diverges to $+\infty$, so does the series $\sum u_n$, by virtue of the comparison Test (Theorem 5-1).

Finally **the series** $\sum_{n=1}^{\infty} a^{1+\frac{1}{2}+\frac{1}{3}+\cdots+\frac{1}{n}}$ **converges for** $0 < a < \frac{1}{e}$ **and diverges for** $a \geq \frac{1}{e}$.

Example 5-12. (The series representation of the number e**)**

The fundamental definition of **the Euler's number** e**, is**

$$e = \lim_{n\to\infty}\left(1+\frac{1}{n}\right)^{n}. \qquad (*)$$

In Example 5-2, we gave the following series representation for the number e,

$$e = 1+\frac{1}{1!}+\frac{1}{2!}+\frac{1}{3!}+\cdots+\frac{1}{n!}+\cdots = \sum_{n=0}^{\infty}\frac{1}{n!}, \quad (0\,! = 1 \text{ definition }). \quad (**)$$

In this Example we shall show that **the infinite series in (**), actually converges to the number** e**, as defined in (*).**

Solution

Let us define the sequences (x_n) and (y_n) with general terms

$$x_n = \left(1+\frac{1}{n}\right)^{n} \text{ and } y_n = 1+\frac{1}{1!}+\frac{1}{2!}+\cdots+\frac{1}{n!}, \text{ respectively.}$$

The first sequence converges to the number e. The second one converges also, (as shown in Example 5-2, for $x = 1$), and furthermore,

$$1+\frac{1}{1!}+\frac{1}{2!}+\frac{1}{3!}+\cdots+\frac{1}{n!}+\cdots < 1+1+\frac{1}{2}+\frac{1}{2^2}+\cdots+\frac{1}{2^n}+\cdots = 1+\frac{1}{1-\frac{1}{2}} = 3.$$

We want to show that

$$\lim x_n = \ln\left(1+\frac{1}{n}\right)^{n} = e = \lim y_n = \sum_{n=0}^{\infty}\frac{1}{n!}.$$

We note that $x_n < y_n$ for $n > 1$. Indeed, by **Newton's Binomial Theorem**,

$$x_n = \left(1+\frac{1}{n}\right)^{n} = 1+1+\frac{n(n-1)}{2!}\frac{1}{n^2}+\cdots+\frac{n(n-1)(n-2)\cdots(n-k+1)}{k!}\frac{1}{n^k}+\cdots+\frac{1}{n^n},$$

or equivalently

$$x_n = 1+\frac{1}{1!}+\frac{1}{2!}\left(1-\frac{1}{n}\right)+\cdots+\frac{1}{k!}\left(1-\frac{1}{n}\right)\left(1-\frac{2}{n}\right)\cdots\left(1-\frac{k-1}{n}\right)+\cdots+$$

$$+\frac{1}{n!}\left(1-\frac{1}{n}\right)\left(1-\frac{2}{n}\right)\cdots\left(1-\frac{n-1}{n}\right), \qquad (***)$$

from which one easily obtains that,

$$x_n < 1+\frac{1}{1!}+\frac{1}{2!}+\cdots+\frac{1}{k!}+\cdots+\frac{1}{n!} \text{ i.e. } \boldsymbol{x_n < y_n} \textbf{ for } \boldsymbol{n} > 1. \ (****)$$

In order to complete our proof, we need the so called **Weierstrass Inequality**, stated below:

If $x_1, x_2, \cdots, x_n$ are numbers positive and less than 1, then

$$(1-x_1)(1-x_2)\cdots(1-x_n) > 1-(x_1+x_2+\cdots x_n).$$

(For a proof see Problem 5-22).

Applying the aforementioned inequality to (***), we have,

$$x_n > 1+\frac{1}{1!}+\frac{1}{2!}\left(1-\frac{1}{n}\right)+\frac{1}{3!}\left(1-\left(\frac{1}{n}+\frac{2}{n}\right)\right)+\cdots+$$

$$+\frac{1}{k!}\left(1-\left(\frac{1}{n}+\frac{2}{n}+\cdots+\frac{k-1}{n}\right)\right)+\cdots+\frac{1}{n!}\left(1-\left(\frac{1}{n}+\frac{2}{n}+\cdots+\frac{n-1}{n}\right)\right), \text{ or}$$

$$x_n > 1+\frac{1}{1!}+\frac{1}{2!}\left(1-\frac{2\cdot 1}{2\cdot n}\right)+\frac{1}{3!}\left(1-\frac{3\cdot 2}{2\cdot n}\right)+\cdots+$$

$$+\frac{1}{k!}\left(1-\frac{k\cdot(k-1)}{2\cdot n}\right)+\cdots+\frac{1}{n!}\left(1-\frac{n\cdot(n-1)}{2\cdot n}\right),$$

or equivalently,

$$x_n > 1+\frac{1}{1!}+\frac{1}{2!}+\frac{1}{3!}+\cdots+\frac{1}{k!}+\cdots+\frac{1}{n!}-$$

$$-\left\{\frac{1}{2!}\frac{2\cdot 1}{2\cdot n}+\frac{1}{3!}\frac{3\cdot 2}{2\cdot n}+\cdots+\frac{1}{k!}\frac{k\cdot(k-1)}{2\cdot n}+\cdots+\frac{1}{n!}\frac{n\cdot(n-1)}{2\cdot n}\right\}, \text{ from which,}$$

$$x_n > y_n-\frac{1}{2\cdot n}\left\{\frac{2\cdot 1}{2!}+\frac{3\cdot 2}{3!}+\cdots+\frac{k\cdot(k-1)}{k!}+\cdots+\frac{n\cdot(n-1)}{n!}\right\}, \text{ and finally,}$$

$$x_n > y_n-\frac{1}{2\cdot n}\left\{1+\frac{1}{1!}+\cdots+\frac{1}{(k-2)!}+\cdots+\frac{1}{(n-2)!}\right\}. \qquad (*****)$$

From (****) and (*****) we easily find that

$$0 < y_n - x_n < \frac{1}{2\cdot n}\left\{1+\frac{1}{1!}+\frac{1}{2!}+\cdots+\frac{1}{(n-2)!}\right\} \text{ for } n > 2, \text{ or}$$

$\mathbf{0 < y_n - x_n < \frac{3}{2 \cdot n}} \Rightarrow \mathbf{\lim(y_n - x_n) = 0}$, and therefore

$\mathbf{\lim y_n = \lim x_n = e}$, and this completes the proof.

Example 5-13.

Show that the number e is irrational. (We know that $2 < e < 3$, see Example 5-12).

Solution

We know that $\mathbf{e = 2 + \frac{1}{2!} + \frac{1}{3!} + \frac{1}{4!} + \cdots + \frac{1}{n!} + \frac{1}{(n+1)!} + \cdots}$ (*)

Let us assume that $\boldsymbol{e}$ is rational. Then e could be expressed as a ratio,

$\boldsymbol{e = \frac{p}{q}}$ **where $\boldsymbol{p}$ and $\boldsymbol{q}$ are positive integers, with $\boldsymbol{q > 1}$.** (**)

From (*) and (**) we have,

$\frac{p}{q} = 2 + \frac{1}{2!} + \frac{1}{3!} + \frac{1}{4!} + \cdots + \frac{1}{n!} + \frac{1}{(n+1)!} + \cdots$ or,

$p(q-1)! - \left\{2 + \frac{1}{2!} + \frac{1}{3!} + \cdots + \frac{1}{q!}\right\} \cdot q! =$

$= \frac{1}{q+1} + \frac{1}{(q+1)(q+2)} + \frac{1}{(q+1)(q+2)(q+3)} + \cdots$, or

$\left|p(q-1)! - \left\{2 + \frac{1}{2!} + \frac{1}{3!} + \cdots + \frac{1}{q!}\right\} \cdot q!\right| < \frac{1}{q+1} + \frac{1}{(q+1)^2} + \frac{1}{(q+1)^3} + \cdots = \frac{1}{q}$

(***)

(Summation of an infinite geometric progression with ratio $\mathbf{0 < \frac{1}{q+1}} < 1$)

But (***) **cannot possibly be true**, since the left side term in (***) **is an integer**, while the **right side term** $\frac{1}{q}$ **is a number** < 1. To avoid the paradox we are forced to assume that the number e cannot be expressed as we have assumed in (**), in other words $\boldsymbol{e}$ **must be an irrational number**, and this completes the proof.

PROBLEMS

5-1) Using arguments similar to the ones applied in the proof of **D' Alembert's Test**, prove Theorem 5-4.

5-2) Test for convergence or divergence,

a) $\sum \frac{1}{15n+3}$ **b)** $\sum \frac{n^{10}}{n!}$ **c)** $\sum \frac{3+\sqrt{n}}{1+n}$ **d)** $\sum n e^{-n}$

5-3) Test for convergence or divergence,

a) $\sum n^5 \left(\frac{4}{7}\right)^n$ **b)** $\sum \frac{(n!)^2}{(2n)!}$ **c)** $\sum \frac{n}{(n+1)(n+5)(n+7)}$ **d)** $\sum \frac{1}{\sqrt{2n^2+5}}$

Hint: For (a) and (b) apply **D' Alembert's Test**, for (c) apply Theorem 5-2, with $\phi_n = \frac{1}{n^2}$ and for (d) apply same Theorem with $\phi_n = \frac{1}{n}$.

(**Answer: a)** Converges, **b)** Converges, **c)** Converges, **d)** Diverges).

5-4) Consider the sequence $u_n = 1 + \frac{1}{2} + \frac{1}{3} + \cdots + \frac{1}{n} - \ln n$ (see Example 5-6), and show that $u_{n+1} < u_n \quad \forall n$.

5-5) For the sequences (u_n) and (s_n) in Example 5-6, show that

$\lim(u_n - s_n) = 0$.

5-6) Work Example 5-7, using **D' Alembert's Test**.

5-7) In Example 5-9 (a), we showed that the series $\sum_{n=2}^{\infty} \frac{1}{2^n} \tan\left(\frac{\pi}{2^n}\right)$ converges. Prove that its sum is $\frac{1}{\pi}$.

Hint: Make use of Example 3-7.

5-8) Using **Cauchy's $n^{\underline{th}}$ Root Test**, show that the series $\sum_{n=2}^{\infty} \frac{1}{(\ln n)^n}$ converges.

5-9) Investigate for convergence the series, $\sum \frac{\ln n}{n^p}$ with $p > 0$,

a) Using **Cauchy's Integral Test,** and

b) Using **Cauchy's Condensation Test.**

(**Answer:** Convergence for $p > 1$, divergence for $0 < p \leq 1$).

5-10) Show that the series $\sum_{n=2}^{\infty} \frac{1}{(\ln n)^{\ln n}}$ converges.

Hint: Show that from a certain point on, $\frac{1}{(\ln n)^{\ln n}} < \frac{1}{n^2}$ and since $\sum \frac{1}{n^2} < \infty$, (**$p$ −series with $p = 2 > 1$**), by dominated convergence, (Theorem 5-1), $\sum_{n=2}^{\infty} \frac{1}{(\ln n)^{\ln n}} < \infty$.

Alternatively, you may apply the **Cauchy's Condensation Test**, to obtain a new series $\sum_{n=2}^{\infty} \left\{\frac{k}{(n \ln k)^{\ln k}}\right\}^n$, where $k > 1$, and then apply the **Cauchy's $n^{\underline{th}}$ Root Test**, etc.

5-11) Test for convergence the series $\sum_{n=2}^{\infty} \frac{1}{n(\ln n)^p}$, where $p > 0$,

a) Using **the Cauchy's Integral Test,** and

b) Using **the Cauchy's Condensation Test.**

(**Answer:** Convergence for $p > 1$, Divergence for $0 < p \leq 1$).

5-12) Test for convergence or divergence the series $\sum \frac{\ln n}{n^3}$.

Hint: You may apply **the Cauchy's Condensation Test,** and then the **D' Alembert's Test,** to show that the series converges.

5-13) Test for convergence or divergence the series $\sum \frac{\ln(n+1)}{n^2}$.

(**Answer:** Converges).

5-14) Test for convergence or divergence,

a) $\sum \frac{3k-2}{2^k}$ **b)** $\sum \frac{1}{\sqrt{k} \cdot \sqrt{k+1}}$

5-15) Show that $\sum \frac{1}{n^n}$ converges.

5-16) Show that $\sum_{n=1}^{\infty} \frac{n^2}{n!} = 2e$.

5-17) Determine the constants A, B, and C, such that
$n^3 = n(n-1)(n-2) + An(n-1) + Bn + C, \quad \forall n \geq 1$, and then show that
$\sum_{n=1}^{\infty} \frac{n^3}{n!}$ converges and find its sum.

(**Answer:** $A = 3, B = 1, C = 0, \ 5e$).

5-18) Show that $\sum \frac{1}{(2n+1)^n}$ converges.

5-19) Test the following series for convergence or divergence,

a) $\sum \frac{n}{\sqrt[5]{n+7}}$ **b)** $\sum n^3 e^{-n}$ **c)** $\sum \frac{\sqrt{n^7+10}}{\sqrt[3]{n^4+n^3}}$.

(**Answer: a)** Diverges, **b)** Converges, **c)** Diverges).

5-20) Consider the sequence (s_n) having general term $s_n = \frac{1!+2!+3!+\cdots+n!}{(n+1)!}$

and show that,

a) The sequence (s_n) is bounded between 0 and 1,

b) The sequence (s_n) is decreasing, and

c) $\lim s_n = 0$.

Hint: $n! < 1! + 2! + 3! + \cdots + n! < n \cdot n!$, therefore $0 < \frac{1}{n+1} < s_n < \frac{n}{n+1} < 1$.
To show that the sequence (s_n) is decreasing, is not difficult. Having established now that $\lim s_n < \infty$, (since (s_n) is decreasing and bounded below), note that $s_{n+1} = \frac{1}{n+2} s_n + \frac{1}{n+2}$, then pass to the limit as $n \to \infty$, etc.

5-21) Test for convergence or divergence,

a) $\sum \frac{2n\sqrt{n}}{(n+1)\sqrt{3n^2+5}}$ **b)** $\sum \frac{(n+2)\sqrt[5]{n^5+2n^2+7}}{(n^4+5n^3+2)\sqrt{n^2+10}}$

Hint: Apply Theorem 5-2, with auxiliary series having general terms $\phi_n = \frac{1}{\sqrt{n}}$ for (a), and $\phi_n = \frac{1}{n^3}$ for (b).

(**Answer: a)** Diverges **b)** Converges).

5-22) Prove the **Weierstrass Inequality**:

If $x_1, x_2, x_3, \cdots, x_n$ are all positive and less than 1, then
$(1-x_1)(1-x_2)(1-x_3)\cdots(1-x_n) > 1-(x_1+x_2+x_3+\cdots+x_n)$.

5-23) Prove that the **Abel's series** $\sum_{n=2}^{\infty} \frac{1}{n \cdot \ln n}$ diverges.

5-24) Test for convergence or divergence,

a) $\sum \frac{(\sin n)^4}{\sqrt[3]{n^5}}$ **b)** $\sum \frac{1}{\ln(n!)}$ **c)** $\sum \frac{\sqrt[n]{n}}{n^2}$ **d)** $\sum \frac{\ln n}{n\sqrt{n}}$

(**Answer: a)** Converges, **b)** Diverges, **c)** Converges **d)** Converges).

5-25) Let $\sum b_n$ be a convergent series with **positive terms.**

i) Prove that $\sum {b_n}^2$ converges.

ii) Show by examples that $\sum \sqrt{b_n}$, may either converge or diverge.

6. Alternating Series.

An **alternating series** is one whose terms are alternately positive and negative, or vice versa. For example, the series

- $\frac{1}{\sqrt{1}} - \frac{1}{\sqrt{2}} + \frac{1}{\sqrt{3}} - \frac{1}{\sqrt{4}} + \frac{1}{\sqrt{5}} - \frac{1}{\sqrt{6}} + \cdots$
- $\frac{1}{1} - \frac{1}{2} + \frac{1}{3} - \frac{1}{4} + \frac{1}{5} - \frac{1}{6} + \cdots$
- $1 - \frac{1}{3} + \left(\frac{1}{3}\right)^2 - \left(\frac{1}{3}\right)^3 + \left(\frac{1}{3}\right)^4 - \left(\frac{1}{3}\right)^5 + \cdots$

are all alternating series. The second series is called **alternating harmonic series**, while the third series is a geometric series with ratio $\left(-\frac{1}{3}\right)$. It can be shown that **the alternating harmonic series converges to the number $\ln 2$**, i.e.

$$\sum_{n=1}^{\infty}(-1)^{n-1}\frac{1}{n} = \frac{1}{1} - \frac{1}{2} + \frac{1}{3} - \frac{1}{4} + \frac{1}{5} - \frac{1}{6} + \cdots = \ln 2 \qquad (6\text{-}1)$$

Any alternating series will be of the form

$$\sum_{n=1}^{\infty}(-1)^{n+1} u_n = u_1 - u_2 + u_3 - u_4 + \cdots, \text{ (where } u_n > 0) \text{ or,} \qquad (6\text{-}2)$$

$$\sum_{n=1}^{\infty}(-1)^{n} u_n = -u_1 + u_2 - u_3 + u_4 - u_5 + \cdots, \text{ (where } u_n > 0). \qquad (6\text{-}3)$$

In order to investigate an alternating series, for convergence or divergence, we use a rather simple test, known as the Leibnitz's Test.

Theorem 6-1 (The Leibnitz's Test).

Let $\sum_{n=1}^{\infty}(-1)^{n+1}u_n$ $(u_n > 0)$, be an alternating series, such that

a) Its terms decrease in absolute value, i.e.

$u_1 > u_2 > u_3 > u_4 > \cdots > u_n > \cdots$, **and**

b) The $\lim u_n = 0$.

Then the alternating series converges to a finite sum S, i.e. $\sum_{n=1}^{\infty}(-1)^{n+1}u_n = S$. Furthermore the sum S lies between any two consecutive partial sums, i.e.

$$\boldsymbol{S_{2n} < S < S_{2n+1}} \qquad (6\text{-}4)$$

By means of (6-4) we **can estimate the sum of the given alternating series.**

Proof: Let $u_1 - u_2 + u_3 - u_4 + \cdots + u_{2n+1} - u_{2n+2} + \cdots$, where $u_k > 0$, $k = 1,2,3,\cdots$, be an alternating series. The sequence of the partial sums of this series, is
$\{S_1 = u_1,\ S_2 = u_1 - u_2,\ S_3 = u_1 - u_2 + u_3,\ S_4 = u_1 - u_2 + u_3 - u_4, \cdots\}$.

Let us now consider the subsequence $\{S_2, S_4, S_6, S_8, \cdots\}$. All the terms of this subsequence **are positive**, because of assumption (a). Also this is **an increasing subsequence**, i.e.

$0 < S_2 < S_4 < S_6 < S_8 < \cdots$. This is easily shown, since

$S_4 - S_2 = u_3 - u_4 > 0,\ S_6 - S_4 = u_5 - u_6 > 0$, etc.

The other **subsequence $\{S_1, S_3, S_5, S_7, \cdots\}$ is positive and decreasing**, i.e.

$0 > S_1 > S_3 > S_5 > \cdots$. Indeed, $S_1 = u_1 > 0,\ S_3 = u_1 - u_2 + u_3 = S_2 + u_3 > 0,\ S_5 = u_1 - u_2 + u_3 - u_4 + u_5 = S_4 + u_5 > 0$, etc.

Also $S_1 - S_3 = -(u_2 - u_3) < 0$, i.e. $S_1 < S_3,\ S_3 - S_5 = -(u_4 - u_5) < 0$, i.e. $S_3 < S_5$, etc. Thus far, we have shown that:

i) The subsequence $\{S_2, S_4, S_6, S_8, \cdots\}$ is positive and increasing, while

ii) The subsequence $\{S_1, S_3, S_5, S_7, \cdots\}$ is positive and decreasing.

In addition $\boldsymbol{S_{2n} < S_{2n+1}},\ \forall \boldsymbol{n}$ since $S_{2n} - S_{2n+1} = -u_{2n+1} < 0$, and

$\boldsymbol{\lim(S_{2n+1} - S_{2n}) = \lim u_{2n+1} = 0}$, because of the assumption (b).

By virtue of **the Theorem of the nested intervals, both subsequences tend to a common limit S, as $n \to \infty$**, and therefore the sequence of the partial sums $\{S_1, S_2, S_3, S_4, S_5, S_6, \cdots\}$ tend to the same limit S, meaning that

$$\boldsymbol{\sum_{n=1}^{\infty}(-1)^{n+1}u_n = u_1 - u_2 + u_3 - u_4 + \cdots = S,}$$

and this completes the proof.

Incidentally, we conclude that the sum S of the alternating series lies between S_{2n} and S_{2n+1}, for any n, a fact which enables us **to estimate the sum of the series.**

Example 6-1.

Show that the alternating harmonic series $1-\frac{1}{2}+\frac{1}{3}-\frac{1}{4}+\frac{1}{5}-\frac{1}{6}+\cdots$, converges.

Solution

The proof follows directly from the Leibnitz's Test, since the $u_n=\frac{1}{n}$ is decreasing and the $\lim u_n=\lim\left(\frac{1}{n}\right)=0$, as $n\to\infty$.

Note: We remind that in contrast, **the harmonic series**

$\mathbf{1+\frac{1}{2}+\frac{1}{3}+\frac{1}{4}+\frac{1}{5}+\frac{1}{6}+\cdots}$**, diverges to** $+\infty$.

Example 6-2.

Show that **the alternating p −series**

$$\sum_{n=1}^{\infty}\frac{(-1)^{n+1}}{n^p}=\frac{1}{1^p}-\frac{1}{2^p}+\frac{1}{3^p}-\frac{1}{4^p}+\frac{1}{5^p}-\frac{1}{6^p}+\cdots,$$

converges for any $p>0$.

Solution

The general term $u_n=\frac{1}{n^p}$ with $p>0$, is decreasing and tends to 0, as $n\to\infty$. According to the Leibnitz's Test, the series converges for all $p>0$.

Note: The p −series (see Example 5-3), converges for $p>1$ and diverges for $0<p\leq 1$.

On the contrary the alternating p −series converges for all $p>0$.

Example 6-3.

Investigate the series $\sum_{n=1}^{\infty}(-1)^{n+1}\frac{n+1}{n}$.

Solution

The terms of the given series decrease in absolute value,

($\frac{n+2}{n+1} < \frac{n+1}{n} \Leftrightarrow n^2 + 2n < n^2 + 2n + 1$), which is true for all n, but the $n^{\underline{th}}$ term **does not** approach zero, as $n \to \infty$, and therefore , by Theorem 4-5, the series diverges.

Example 6-4.

Test for convergence or divergence, the series $\sum_{n=1}^{\infty} \sin\left\{\left(n + \frac{k}{n}\right)\pi\right\}$, where $k > 0$.

Solution

$\sum_{n=1}^{\infty} \sin\left\{\left(n + \frac{k}{n}\right)\pi\right\} = \sum_{n=1}^{\infty}\left\{\sin(n\pi)\cos\left(\frac{k\pi}{n}\right) + \cos(n\pi)\sin\left(\frac{k\pi}{n}\right)\right\} =$
$\sum_{n=1}^{\infty}(-1)^n \sin\left(\frac{k\pi}{n}\right)$.

given that $\mathbf{sin(n\pi) = 0}$ and $\mathbf{cos(n\pi) = (-1)^n}$.

Since $k > 0$, from a certain point on, this will be an alternating series, with terms decreasing in absolute value, and approaching zero, as $n \to \infty$, therefore by the Leibnitz's Test, the series converges.

PROBLEMS

6-1) Show that the series in Example 6-4, converges for every $k \in \mathbb{R}$, (and not just for $k > 0$, as it was assumed in the Example).

6-2) Test the following alternating series for convergence,

a) $\sum \frac{(-1)^{n+1}}{\sqrt{n}}$ **b)** $\sum \frac{(-1)^{n+1}}{\ln n}$ **c)** $\sum (-1)^n n^{-3}$ **d)** $\sum \frac{(-1)^n 2^n}{n!}$.

6-3) Investigate the series,

a) $\sum \frac{(-1)^k \ln k}{k}$ **b)** $\sum \frac{(-1)^k}{4k+5}$

(**Answer: a)** Converges **b)** Converges).

6-4) Making use of **Leibnitz's Test,** show that the following series converge,

a) $\sum \frac{\cos(n\pi)}{n^3}$ **b)** $\sum(-1)^{n+1}\{\sqrt{n+2}-\sqrt{n+1}\}$.

6-5) Investigate the series,

a) $\sum(-1)^n\left(1+\frac{1}{n}\right)^{-n}$ **b)** $\sum_{n=3}^{\infty}\frac{(-1)^{n+1}}{\ln(\ln n)}$.

(**Answer: a)** Diverges, **b)** Converges).

6-6) Show that both series diverge,

a) $\sum(-1)^{n+1}\ln\left(\frac{1}{n^2}\right)$ **b)** $\sum(-1)^{n+1}\ln\left(\sqrt[n]{n}\right)$.

6-7) Determine whether the given series converge or diverge,

a) $\sum(-1)^{n+1}\frac{\ln n}{n}$ **b)** $\sum\frac{(-1)^{n+1}\cos(n\pi)}{n}$ **c)** $\sum(-1)^n\cos\left(\frac{\pi}{n^3}\right)$

d) $\sum(-1)^n\frac{\sqrt[n]{n+20}}{n+3}$.

(**Answer: a)** Converges, **b)** Diverges, **c)** Diverges, **d)** Converges).

7. Absolute and Conditional Convergence.

Definition 7-1: A series $\sum \boldsymbol{u_n}$ is said to be absolutely convergent, if the series of the absolute values of its terms is convergent, i.e. if the series

$$\sum|\boldsymbol{u_n}| = |\boldsymbol{u_1}| + |\boldsymbol{u_2}| + |\boldsymbol{u_3}| + |\boldsymbol{u_4}| + \cdots + |\boldsymbol{u_n}| + \cdots \qquad (7\text{-}1)$$

is convergent, (($\sum|\boldsymbol{u_n}| < \infty$).

Definition 7-2: If a series $\sum \boldsymbol{b_n}$ converges, but the series of the absolute values of its terms diverges,(i.e. if $\sum \boldsymbol{b_n} < \infty$ but $\sum|\boldsymbol{b_n}| = +\infty$),then the series $\sum \boldsymbol{b_n}$ is said to be conditionally convergent.

For example the series

$$\sum_{n=1}^{\infty} \frac{(-1)^{n+1}}{n^2} = \frac{1}{1^2} - \frac{1}{2^2} + \frac{1}{3^2} - \frac{1}{4^2} + \frac{1}{5^2} - \frac{1}{6^2} + \cdots$$

is **absolutely convergent**, since the series of the absolute values of its terms,

$$\sum_{n=1}^{\infty} \left|\frac{(-1)^n}{n^2}\right| = \frac{1}{1^2} + \frac{1}{2^2} + \frac{1}{3^2} + \frac{1}{4^2} + \frac{1}{5^2} + \frac{1}{6^2} + \cdots$$

is convergent,(**it is a $\boldsymbol{p}$ –series with $\boldsymbol{p} = \mathbf{2} > 1$**, see Example 5-3).

On the other hand, the series

$$\sum_{n=1}^{\infty} \frac{(-1)^{n+1}}{n} = \frac{1}{1} - \frac{1}{2} + \frac{1}{3} - \frac{1}{4} + \frac{1}{5} - \frac{1}{6} + \cdots$$

is **conditionally convergent**, since the given series converges to the number $\mathbf{ln\,2}$, (see equation 6-1), but the series of its absolute values,

$$\sum_{n=1}^{\infty} \left|\frac{(-1)^{n+1}}{n}\right| = \sum_{n=1}^{\infty} \frac{1}{n} = \frac{1}{1} + \frac{1}{2} + \frac{1}{3} + \frac{1}{4} + \frac{1}{5} + \frac{1}{6} + \cdots$$

diverges to $+\infty$, (**Harmonic Series**, see Theorem 4-6).

The concept of **absolute convergence** is important, because of the following Theorem.

Theorem 7-1.

Every absolutely convergent series is itself convergent, i.e. if the series $\sum|u_n|$ converges, then $\sum u_n$ converges as well, (not to the same sum, but it converges).

Proof: Let us assume that $\sum|u_n|$ converges. It is a trivial matter to show that,

$0 \leq |u_n| - u_n \leq 2|u_n|$, and (*)

$0 \leq |u_n| + u_n \leq 2|u_n|$.

(**)

By virtue of Theorem 5-1, (the Comparison Test), both series,

$\sum(|u_n| - u_n)$, and $\sum(|u_n| + u_n)$

are convergent, therefore their difference,

$\sum\{(|u_n| + u_n) - (|u_n| - u_n)\} = 2u_n$ (***)

will also be convergent, and finally **the series $\sum u_n$ will be convergent as well**, (see Theorem 4-1).

Remarks:

- Given a series **with arbitrary terms**, Theorem 7-1 shows that **if the series is absolutely convergent**, then the original series will be convergent as well. For example the series

$$\frac{1}{1^3} + \frac{1}{2^3} - \frac{1}{3^3} - \frac{1}{4^3} - \frac{1}{5^3} + \frac{1}{6^3} + \frac{1}{7^3} - \frac{1}{8^3} - \frac{1}{9^3} - \frac{1}{10^3} + \cdots$$

(Two positive terms followed by three negative ones), converges since the given series is absolutely convergent,(**it is a p −series with $p = 3 > 1$**).

- **If the series $\sum|u_n|$ diverges, the series itself $\sum u_n$, may be either convergent or divergent**. For example let us consider the two series,

$$\frac{1}{1} - \frac{1}{2} + \frac{1}{3} - \frac{1}{4} + \frac{1}{5} - \frac{1}{6} + \frac{1}{7} - \frac{1}{8} + \cdots, \text{ and} \quad (*)$$

$$\frac{1}{1} + \frac{1}{2} + \frac{1}{3} + \frac{1}{4} + \frac{1}{5} + \frac{1}{6} + \frac{1}{7} + \frac{1}{8} + \cdots.$$

(**)

Both series **are absolutely divergent**, i.e. $\sum|u_n| = +\infty$, however the first series is convergent (to $\ln 2$), while the second one is divergent to $+\infty$.

- Given a series $\sum u_n$ with **terms of arbitrary signs** (positive and negative), Theorem 7-1 shows that **there are exactly three possibilities:**
  1) The series is absolutely convergent,
  2) The series is conditionally convergent,
  3) The series is divergent.

Suppose now that we wish to **investigate the nature of a given series $\sum u_n$ with arbitrary terms**.

1) We may start by considering the series $\sum|u_n|$. If the latter series converges, i.e. **if the given series is absolutely convergent, then by Theorem 7-1, the original series $\sum u_n$ will be convergent as well**. We note that **since the terms of $\sum|u_n|$ are positive, all Theorems and Techniques developed in Chapter 5 (Convergence criteria for series with positive terms), can be applied**. In particular, we mention the following two Theorems, which are used quite often in practice.

Theorem 7-2 (The Comparison Test).

If $|u_n| \leq c|b_n|$ where c is a fixed positive constant and the series $\sum b_n$ is absolutely convergent, then the series $\sum u_n$ is convergent as well.

For a proof see Problem 7-21.

Theorem 7-3 (The D' Alembert's Test).

Let $\sum u_n$ be a series with arbitrary terms, and let also $r = \lim \frac{|u_{n+1}|}{|u_n|}$. Then

a) If $0 \leq r < 1$, the series $\sum u_n$ converges absolutely, and therefore, by Theorem 7-1, the series $\sum u_n$ converges as well,

b) If $r > 1$, the series $\sum u_n$ diverges,

c) If $r = 1$ the Test is inconclusive, i.e. the series may be either convergent or divergent.

For a proof see Problem 7-22.

2) If $\sum|u_n|$ diverges, ($\sum|u_n| = +\infty$), the original series $\sum u_n$ could be either convergent or divergent. Its nature should be revealed by means of some other methods.

Definition 7-3: Given an arbitrary series, $\sum u_n = u_1 + u_2 + u_3 + u_4 + u_5 + \cdots$ we may form a new series by **listing the terms in a different order**, for example,

$u_1 + u_5 + u_7 + u_2 + u_3 + u_4 + u_8 + u_6 + \cdots$ or

$u_3 + u_2 + u_1 + u_6 + u_5 + u_4 + u_9 + u_8 + \cdots$, etc.

Such a new series is called **a rearrangement** of the original series.

An essential difference between **absolute convergence and conditional convergence** is revealed in the following two Theorems.

Theorem 7-4.

Let $\sum u_n$ be an absolutely convergent series. Then every rearrangement of this series is also convergent and has the same sum with the original series.

For a proof see Example 7-6.

Next Theorem, which we state without proof, deals with **rearrangements of conditionally convergent series.**

Theorem 7-5.

Let $\sum u_n$ be a conditionally convergent series. Then,

a) The series has a rearrangement diverging to $+\infty$,

b) The series has another rearrangement diverging to $-\infty$,

c) If x is any real number, there exists a rearrangement of the original series, converging to the number x.

So in general, we conclude that **rearrangement of the terms, in a conditionally convergent series, alters the sum of the series,** (see Example 4-3 and Problem 4-5).

However, in **a convergent series with positive terms or in an absolutely convergent series, we can arbitrarily rearrange its terms, without affecting its sum.**

In summary, we note that **absolutely convergent series, like ordinary finite sums, possess the commutative property, i.e. any rearrangement of the terms does not affect its sum.**

Another characteristic property of absolutely convergent series is related to the operation **of multiplication** of the series.

Definition 7-4: The **Cauchy product** of two convergent series,

$S_1 = u_1 + u_2 + u_3 + u_4 + u_5 + \cdots + u_n + \cdots$, and (*)

$S_2 = b_1 + b_2 + b_3 + b_4 + b_5 + \cdots + b_n + \cdots$, (**)

is defined as,

$$(\boldsymbol{u_1 b_1}) + (\boldsymbol{u_1 b_2} + \boldsymbol{u_2 b_1}) + (\boldsymbol{u_1 b_3} + \boldsymbol{u_2 b_2} + \boldsymbol{u_3 b_1}) + \cdots + (\boldsymbol{u_1 b_n} + \boldsymbol{u_2 b_{n-1}} + \cdots + \boldsymbol{u_{n-1} b_2} + \boldsymbol{u_n b_1}) + \cdots$$

or, in compact form as,

$\sum_{n=1}^{\infty} \boldsymbol{a_n}$, **where** $\boldsymbol{a_n} = \sum_{k=1}^{n} \boldsymbol{u_k b_{n+1-k}}$ (7-2)

The following Theorem, which we state without proof, reveals another important property, related to **the multiplication of absolutely convergent series.**

Theorem 7-6.

Let $\sum_{n=1}^{\infty} \boldsymbol{u_n} = \boldsymbol{S_1}$ **and** $\sum_{n=1}^{\infty} \boldsymbol{b_n} = \boldsymbol{S_2}$ **be two absolutely convergent series. Then their Cauchy product, as defined in (7-2), is also an absolutely convergent series, with sum** $\boldsymbol{S} = \boldsymbol{S_1} \cdot \boldsymbol{S_2}$**, i.e.**

$\boldsymbol{S} = \sum_{n=1}^{\infty} \boldsymbol{a_n} = \boldsymbol{S_1} \cdot \boldsymbol{S_2}$, **where** $\boldsymbol{a_n} = \sum_{k=1}^{n} \boldsymbol{u_k b_{n+1-k}}$ (7-3)

Example 7-1.

a) Show that the following series converges,

$$\frac{1}{1^2} - \frac{1}{2^2} + \frac{1}{3^2} - \frac{1}{4^2} + \frac{1}{5^2} - \frac{1}{6^2} + \frac{1}{7^2} - \frac{1}{8^2} + \cdots$$

b) Given that $\sum_{n=1}^{\infty}\frac{1}{n^2}=\frac{\pi^2}{6}$, show that $\sum_{n=1}^{\infty}\frac{1}{(2n-1)^2}=\frac{\pi^2}{8}$ and then find the sum of the series in (a).

Solution

a) The given series is absolutely convergent, (**it is a p –series with $p=2>1$**), therefore , by virtue of Theorem 7-1, the given series converges, say to a finite number S, which is to be found in part (b).

b) It is given that

$$\sum_{n=1}^{\infty}\frac{1}{n^2}=\frac{\pi^2}{6} \text{ i.e. } \frac{1}{1^2}+\frac{1}{2^2}+\frac{1}{3^2}+\frac{1}{4^2}+\frac{1}{5^2}+\frac{1}{6^2}+\cdots=\frac{\pi^2}{6}$$

Since this is an absolutely convergent series,(obviously since **its terms are all positive**), any rearrangement of this series will converge to the same sum $\frac{\pi^2}{6}$, i.e.

$$\frac{1}{1^2}+\frac{1}{3^2}+\frac{1}{5^2}+\cdots+\frac{1}{2^2}+\frac{1}{4^2}+\frac{1}{6^2}+\cdots=\frac{\pi^2}{6} \text{ or,}$$

$$\frac{1}{1^2}+\frac{1}{3^2}+\frac{1}{5^2}+\cdots+\frac{1}{2^2}\left(\frac{1}{1^2}+\frac{1}{2^2}+\frac{1}{3^2}+\cdots\right)=\frac{\pi^2}{6} \text{ or,}$$

$$\sum_{n=1}^{\infty}\frac{1}{(2n-1)^2}+\frac{1}{4}\cdot\frac{\pi^2}{6}=\frac{\pi^2}{6}\Rightarrow\sum_{n=1}^{\infty}\frac{\mathbf{1}}{(\mathbf{2n-1})^2}=\frac{\mathbf{1}}{\mathbf{1^2}}+\frac{\mathbf{1}}{\mathbf{3^2}}+\frac{\mathbf{1}}{\mathbf{5^2}}+\cdots=\frac{\boldsymbol{\pi}^2}{\mathbf{8}}.$$

Again by virtue of Theorem 7-4 every rearrangement of the given in (a), absolutely convergent series, converges to the same sum S, as the original series, i.e.

$$S=\frac{1}{1^2}+\frac{1}{3^2}+\frac{1}{5^2}+\cdots-\frac{1}{2^2}\left(\frac{1}{1^2}+\frac{1}{2^2}+\frac{1}{3^2}+\cdots\right)\Rightarrow \boldsymbol{S}=\frac{\boldsymbol{\pi}^2}{\mathbf{8}}-\frac{\mathbf{1}}{\mathbf{4}}\cdot\frac{\boldsymbol{\pi}^2}{\mathbf{6}}=\frac{\boldsymbol{\pi}^2}{\mathbf{12}}.$$

Example 7-2.

If $\sum u_n$ is an absolutely convergent series, then the series

$\sum \boldsymbol{u_n}\sin(\boldsymbol{f(n)})$ and $\sum \boldsymbol{u_n}\cos(\boldsymbol{g(n)})$

where $f(n)$ and $g(n)$ are arbitrary functions of n, are also absolutely convergent.

Solution

Since $|u_n \sin(f(n))| \leq |u_n|$ and $\sum |u_n| < \infty$ (by assumption), by virtue of Theorem 7-2, (with $c = 1$), the series $\sum u_n \sin(f(n))$ will be absolutely convergent as well.

A similar proof applies for the series $\sum u_n \cos(g(n))$.

Note: Depending on the particular form of u_n, $f(n)$ and $g(n)$, it is possible, in some cases to obtain the sum of the corresponding series **in closed form**. For example, in Chapter 9, Problem 9-4, it is shown that,

$$\boldsymbol{\sin x - \frac{\sin(3x)}{3!} + \frac{\sin(5x)}{5!} - \frac{\sin(7x)}{7!} + \cdots = \cos(\cos x) \cdot \sinh(\sin x)},$$

where x is any real number.(**The** $\boldsymbol{\sinh x}$ **is the hyperbolic sine function, defined as** $\boldsymbol{\sinh x = \frac{e^x - e^{-x}}{2}}$). For a brief introduction to the **Hyperbolic Functions**, see next Chapter 8.

Example 7-3.

Show that the series $\sum_{k=2}^{\infty} = (-1)^k \frac{1}{\ln k}$ is conditionally convergent.

Solution

Since $\frac{1}{\ln 2} > \frac{1}{\ln 3} > \frac{1}{\ln 4} > \frac{1}{\ln 5} > \cdots$, and $\lim \left(\frac{1}{\ln n}\right) = 0$,

application of the **Leibnitz's Test** shows that **the given alternating series converges**. However, the given series **is not absolutely convergent, since** $\boldsymbol{\sum_{k=2}^{\infty} \frac{1}{\ln k} = +\infty}$.

Therefore the series **is conditionally convergent**.

Note: To prove that $\sum_{k=2}^{\infty} \frac{1}{\ln k} = +\infty$, show first that $\frac{1}{k} < \frac{1}{\ln k}$, $k = 2,3,4,\cdots$, and then make use of Theorem 5-1(b), and the fact that the Harmonic Series diverges to $+\infty$.

Example 7-4.

Find all the values of x, for which the following series converges,

$\sum_{n=1}^{\infty} \frac{x^n}{n} = \frac{x}{1} + \frac{x^2}{2} + \frac{x^3}{3} + \frac{x^4}{4} + \frac{x^5}{5} + \cdots.$

Solution

Let us consider the series of the absolute values of the terms of the series $\sum_{n=1}^{\infty} \frac{|x|^n}{n} = \frac{|x|^1}{1} + \frac{|x|^2}{2} + \frac{|x|^3}{3} + \frac{|x|^4}{4} + \frac{|x|^5}{5} + \cdots$

and apply Theorem 7-3 (**D' Alembert's Test**).

Let $\boldsymbol{r} = \mathbf{lim} \frac{|\boldsymbol{x}|^{\boldsymbol{n}+1} \div (\boldsymbol{n}+1)}{|\boldsymbol{x}|^{\boldsymbol{n}} \div \boldsymbol{n}} = |\boldsymbol{x}| \cdot \mathbf{lim} \frac{\boldsymbol{n}+1}{\boldsymbol{n}} = |\boldsymbol{x}| \cdot \mathbf{1} = |\boldsymbol{x}|.$

a) If $r = |x| < 1$, i.e. if $-\mathbf{1} < \boldsymbol{x} < \mathbf{1}$, the given series converges, while for $r = |x| > 1$, i.e. for $-\infty < \boldsymbol{x} < -\mathbf{1}$, or $\mathbf{1} < \boldsymbol{x} < +\infty$, the series diverges.

b) At the points $x = \pm 1$, where $|x| = 1$, special attention is needed, **and convergence at these points, has to be examined separately**.

At $x = 1$, the series becomes, $\frac{1}{1} + \frac{1}{2} + \frac{1}{3} + \frac{1}{4} + \frac{1}{5} + \cdots = +\infty$, (Diverges), while

At $x = -1$, the series becomes, $-\frac{1}{1} + \frac{1}{2} - \frac{1}{3} + \frac{1}{4} - \frac{1}{5} + \cdots = -\ln 2$, (Converges, see (6-1)).

Finally, **the given series converges for** $-\mathbf{1} \leq \boldsymbol{x} < \mathbf{1}$.

Example 7-5.

Consider the series $\sum u_n$ where $u_n = 1 - \cos\left(\frac{2}{n^x}\right)$, and find the real values of x, for which the series converges.

Solution

Making use of the well known Trigonometric Identity

$(1 - \cos x) = 2(\sin(\frac{x}{2}))^2$, the series

$\sum u_n = 2\sum\left(\sin(\frac{1}{n^x})\right)^2.$ (*)

Let us now consider the series $\sum \left(\frac{1}{n^x}\right)^2 = \sum \frac{1}{n^{2x}}$ (**)

as **an auxiliary series**, and find the limit,

$$\lim \frac{\left(\sin\left(\frac{1}{n^x}\right)\right)^2}{\left(\frac{1}{n^x}\right)^2} = \lim \left(\frac{\sin\left(\frac{1}{n^x}\right)}{\frac{1}{n^x}}\right)^2 = 1, \text{ provided that } x > 0.$$

(This is so since $\lim_{y \to 0} \frac{\sin y}{y} = 1$. In our case consider $\boldsymbol{y = \frac{1}{n^x}}$).

By virtue of Theorem 5-2, (**The limit Comparison Test**), **the series (*) and (**) are of the same nature**, and since the series (**) converges for $2x > 1$, i.e. for $x > \frac{1}{2}$, (see Example 5-3, the p –series), we conclude that the series (*), i.e. **the series $\sum u_n$ converges for $x > \frac{1}{2}$ as well.**

Example 7-6.

Prove Theorem 7-4.

Solution

a) We shall first prove the following:

Any possible rearrangement of a convergent series with positive terms, does not affect its sum.

Let $\boldsymbol{u_1 + u_2 + u_3 + \cdots + u_n + \cdots = A}$ (*)

be a convergent series with positive terms.

Let also $\boldsymbol{b_1 + b_2 + b_3 + \cdots + b_n + \cdots}$ (**)

be **any rearrangement of the series (*)**. This means that every term in (*) appears in another position in (**), **and conversely** every term in (**) appears in (*), in a different position.

Let $\overline{s_m} = b_1 + b_2 + b_3 + \cdots + b_m$ (***)

be the sum of the first m terms, of the series in (**).

Since all these terms appear also in (*), in a different position, if we choose n to be quite large, the m first terms in (**), i.e. the terms $b_1, b_2, b_3, \cdots, b_m$ will be among the first n terms of the series (*), and therefore,

$$\overline{s_m} < s_n = u_1 + u_2 + u_3 + \cdots + u_n < A. \qquad (****)$$

As $m \to \infty$, the sequence of the partial sums $(\overline{s_m})$ is increasing and bounded above from the number A, meaning that **the sequence $(\overline{s_m})$ tends to a limit $B \leq A$**, i.e.

$$\lim(\overline{s_m}) = b_1 + b_2 + b_3 + \cdots + b_n + \cdots = B \leq A. \qquad (*****)$$

At the same time, the original series (*), can be obtained from (**), **by rearranging the terms of the latter**, and applying the same arguments, as before, we can show that $A \leq B$, which when combined with (*****), yields that $A = B$, i.e. $\sum u_n = \sum b_n$, and this completes the proof.

b) We shall now prove Theorem 7-4.

In Theorem 7-1, it was shown that **an absolutely convergent series can be expressed as a difference of two positive convergent series,**(see (*), (**) and (***) in the proof of Theorem 7-1), therefore any rearrangement of the terms of the original series, reduces to a rearrangement of the terms of these two positive series, which by virtue of the part (a) above, **does not affect their sums.**

PROBLEMS

7-1) If $\sum {u_n}^2 < \infty$ and $\sum {b_n}^2 < \infty$, show that the series $\sum(u_n b_n)$ is absolutely convergent.

Hint: $|u_n b_n| \leq \frac{1}{2}\left({u_n}^2 + {b_n}^2\right)$

7-2) Test for conditional and absolute convergence and following series,

a) $\sum(-1)^{n+1}\sin(\frac{\pi}{n})$ **b)** $\sum \frac{\cos(n\pi)}{\sqrt{n+10}}$.

7-3) Consider the series $\sum_{n=2}^{\infty}\left(\frac{1}{\sqrt{n}-1}-\frac{1}{\sqrt{n}+1}\right)$, and show that it diverges. Is the given series absolutely convergent?

(**Answer:** No).

7-4) Show that the series $\sum \frac{n!x^n}{n^n}$ converges for $-e \le x < e$.

Hint: Application of Theorem 7-3, shows that for $|x| < e$, the series converges. However at the end points, $x = \pm e$, the Test fails. We need to examine these points separately. At $x = -e$, the series becomes an alternating series, which by the Leibnitz's Test converges. At $x = e$, the series becomes $\sum u_n$ where $u_n = \frac{n!e^n}{n^n}$. Show that $e = u_1 < u_2 < u_3 < \cdots$, and since $\lim u_n \ne 0$, the series diverges.

7-5) Find the values of x for which the following series converge,

a) $\sum \frac{(x+2)^n}{n}$ **b)** $\sum \frac{(2x+4)^n}{n!}$ **c)** $\sum \frac{(x-1)^n}{(n+1)^2}$ **d)** $\sum (-1)^{n+1}\frac{\sin(nx+x^2)}{\sqrt{n^3}}$

Hint: See Example 7-4, for (a), (b), and (c), and Example 7-2 for (d).

(**Answer:** **a)** $-3 \le x < -1$, **b)** $-\infty < x < +\infty$,

c) $0 \le x \le 2$, **d)** $-\infty < x < +\infty$).

7-6) Find the convergence set for the given series,

a) $\sum \frac{2^n}{n}\left(\frac{x-3}{x+2}\right)^n$ **b)** $\sum \frac{(x-n)^n}{n+1}$ **c)** $\sum (-1)^{n+1}\frac{1}{2n+x^2}$

d) $\sum \frac{\sin(n^4+x)}{n^3}$ **e)** $\sum \frac{\ln n}{n}\left(\frac{2x+1}{2x+3}\right)^n$ **f)** $\sum \frac{1}{nx^n}$.

Hint: Special attention should be given at the end points.

7-7) Find the convergence set for the given series,

a) $\sum \frac{(x^2-3)^n}{5^n}$ **b)** $\sum (-1)^n x^{\left(\frac{n^2}{n+1}\right)}$ **c)** $\sum \frac{n^3+2}{(n+1)!}x^n$ **d)** $\sum (-1)^n \frac{x^n}{2n+3\sqrt{n}}$

(**Answer:** **a)** $0 \le x < \sqrt{8}$, **b)** $-1 < x < 1$,

c) $-\infty < x < +\infty$, **d)** $-1 < x \le 1$).

7-8) Show that the series $\sum_{n=1}^{\infty} x^n \tan\left(\frac{x}{2^n}\right)$ converges for $-2 < x < 2$.

Hint: The $\lim_{y\to 0} \frac{\tan y}{y} = 1$.

7-9) If $x = \frac{1}{2}$, find the sum of the series in Problem 7-8, (see Example 3-7).

(Answer: $\frac{1}{x} - \cot x$).

7-10) Show that the series $1 - \frac{1}{\sqrt[3]{2}} + \frac{1}{\sqrt[3]{3}} - \frac{1}{\sqrt[3]{4}} + \cdots$, is conditionally convergent.

7-11) Show that the following series are convergent for all values of x,

a) $\sum \frac{\sin(n\pi x)}{\sqrt{n^3}}$ **b)** $\sum \frac{\cos(\sin(2nx))}{n^2}$

c) $\sum \frac{\sin(nx)+\cos(\sqrt{n}x^2)}{\sqrt[5]{n^7}}$ **d)** $\sum (-1)^n \frac{\cos\left(\frac{x}{n}\right)}{\sqrt[3]{n^8}}$

7-12) Show that the series $\sum \frac{x^n}{3n+x^{2n}}$ converges $\forall x \in \mathbb{R} - \{1\}$.

7-13) Investigate the following series for convergence or divergence,

a) $\sum \frac{n\cos(n\pi)}{n^3+7}$ **b)** $\sum (-1)^{n+1} \frac{1}{\sqrt{n+1}-\sqrt{n}}$ **c)** $\sum (-1)^{n+1} \frac{1}{\sqrt{n+1}+\sqrt{n}}$

d) $\sum \left(-\frac{1}{\ln n^2}\right)^n$ **e)** $\sum (-1)^{n+1} \left(\frac{n+3}{2n+5}\right)^n$ **f)** $\sum (-1)^n \frac{\ln n}{n}$

(Answer: a) Absolutely convergent, **b)** Divergent, **c)** Conditionally convergent, **d)** Absolutely convergent, **e)** Absolutely convergent **f)** Conditionally convergent).

7-14) Show that if $\sum u_n$ is absolutely convergent, then the series $\sum {u_n}^2$ is convergent. Note that **the converse proposition is not necessarily true**. Give an Example of a series, such that $\sum {u_n}^2 < \infty$, but $\sum |u_n| = +\infty$.

7-15) Show that the following series converges, and find its sum,

$$\frac{1}{1^2} - \frac{1}{2^2} - \frac{1}{4^2} - \frac{1}{6^2} + \frac{1}{3^2} - \frac{1}{8^2} - \frac{1}{10^2} - \frac{1}{12^2} + \frac{1}{5^2} - \frac{1}{14^2} - \frac{1}{16^2} - \frac{1}{18^2} + \cdots.$$

(each positive term is followed by three negative ones).

Hint: See Example 7-1(a), and make use of Theorem 7-4.

(Answer: $\frac{\pi^2}{12}$).

7-16) It is known that $\sum_{n=0}^{\infty} x^n = \frac{1}{1-x}$ where $|x| < 1$, (see Theorem 2-1). Taking the **Cauchy product** of this series by itself, show that

$$\sum_{n=0}^{\infty}(n+1)x^n = \frac{1}{(1-x)^2}, \quad (|x| < 1).$$

Hint: Apply Theorem 7-6.

7-17) In Example 5-2, we introduced **the exponential function** $\boldsymbol{e^x}$, by means of the infinite series,

$$\boldsymbol{e^x = \sum_{n=0}^{\infty}\frac{x^n}{n!} = 1 + \frac{x}{1!} + \frac{x^2}{2!} + \frac{x^3}{3!} + \frac{x^4}{4!} + \cdots}, \qquad -\infty < x < \infty.$$

a) Show that this series **is absolutely convergent, for every real value of** $\boldsymbol{x}$. (Make use of Theorem 7-3).

b) If x and y are any two real numbers, show with the aid of Theorem 7-6, that $\boldsymbol{e^x \cdot e^y = e^{x+y}}$, and then show that $\boldsymbol{e^x \div e^y = e^{x-y}}$.

7-18) If a series with positive terms diverges to $+\infty$, then any rearrangement of the original series diverges to $+\infty$ as well.

7-19) Show that the series $\sum_{n=1}^{\infty} n^2 x^{n-1}$ converges for $|x| < 1$, and find its sum.

(Answer: $\frac{1+x}{(1-x)^3}$).

7-20) For which values of $p > 0$, the following series is convergent?

$$\frac{1}{2^p-1} - \frac{1}{2^p+1} + \frac{1}{3^p-1} - \frac{1}{3^p+1} + \cdots + \frac{1}{n^p-1} - \frac{1}{n^p+1} + \cdots.$$

7-21) Prove Theorem 7-2.

7-22) Prove Theorem 7-3.

8. Complex Numbers and Hyperbolic Functions. (A Brief Review)

In this chapter we give a brief, **systematic** summary **of complex numbers and hyperbolic functions**, needed for the further development of the Theory of Series, from real to complex numbers.

A) Complex Numbers.

It is known that every number of the form

$\boldsymbol{z = x + iy}$**, where** $\boldsymbol{x}$ **and** $\boldsymbol{y}$ **are real numbers, and** $\boldsymbol{i = \sqrt{-1}}$**, (or** $\boldsymbol{i^2 = -1}$**)**

is called **a complex number**. We shall use the symbol $\mathbb{C}$ to denote the set of complex numbers.

The real number x is called **the real part of** $\boldsymbol{z}$, while the real number y is called the **imaginary part of** $\boldsymbol{z}$, and we write,

$\boldsymbol{x = \Re(z)}$ **and** $\boldsymbol{y = \Im(z)}$.

The number $\boldsymbol{i = \sqrt{-1}}$ **is called the imaginary unit.**

Two complex numbers $z = x + iy, (x \in \mathbb{R}, y \in \mathbb{R})$ and $w = a + ib, (a \in \mathbb{R}, b \in \mathbb{R})$, are said to be **equal if and only if they have equal real and imaginary parts**, i.e.

$$\boldsymbol{z = w \Leftrightarrow x + iy = a + ib \Leftrightarrow \{x = a \ \ and \ \ y = b\}}. \qquad (8\text{-}1)$$

Inequalities between complex numbers **are not defined**.

If $z = x + iy$ is a complex number, **the modulus or absolute value of** $\boldsymbol{z}$, is a **non negative** real number, denoted by $|\boldsymbol{z}|$ **or** $\boldsymbol{r}$, defined as,

$$\boldsymbol{z = x + iy\,, \qquad r = |z| = \sqrt{x^2 + y^2} \geq 0}. \qquad (8\text{-}2)$$

If $|z| = 0$, then $x = 0$ and $y = 0$, (why?), meaning that the **only complex number with modulus 0, is the complex number** $\boldsymbol{0 + i0}$.

In all other cases $|\boldsymbol{z}| > 0$.

The complex number $z = x - iy$, denoted by $\bar{z}$, is called the **complex conjugate of $z = x + iy$**, i.e.

If $z = x + iy$ then $\bar{z} = x - iy$. (8-3)

We note that,

$$z\bar{z} = |z|^2 \text{ and } \overline{(\bar{z})} = z. \tag{8-4}$$

The following properties are easily proved:

$$\overline{z_1 + z_2 + \cdots + z_n} = \overline{z_1} + \overline{z_2} + \cdots + \overline{z_n} \tag{8-5}$$

$$\overline{z_1 z_2 \cdots z_n} = \overline{z_1 z_2} \cdots \overline{z_n} \text{ and } \overline{\left(\frac{z_1}{z_2}\right)} = \frac{\overline{z_1}}{\overline{z_2}} \tag{8-6}$$

$$|z_1 + z_2 + \cdots + z_n| \leq |z_1| + |z_2| + \cdots + |z_n| \tag{8-7}$$

$$|z_1 z_2 \cdots z_n| = |z_1||z_2| \cdots |z_n| \text{ and } \left|\frac{z_1}{z_2}\right| = \frac{|z_1|}{|z_2|} \tag{8-8}$$

If $z_1 = z_2 = \cdots z_n = z$, then from (8-6) and (8-8), we have,

$$\overline{z^n} = (\bar{z})^n \text{ and } |z^n| = |z|^n \tag{8-9}$$

The expression $x + iy$ is called **the Cartesian expression of the complex number z**, and is represented by a vector in the Cartesian x-y plane, as shown in Fig.8-1.

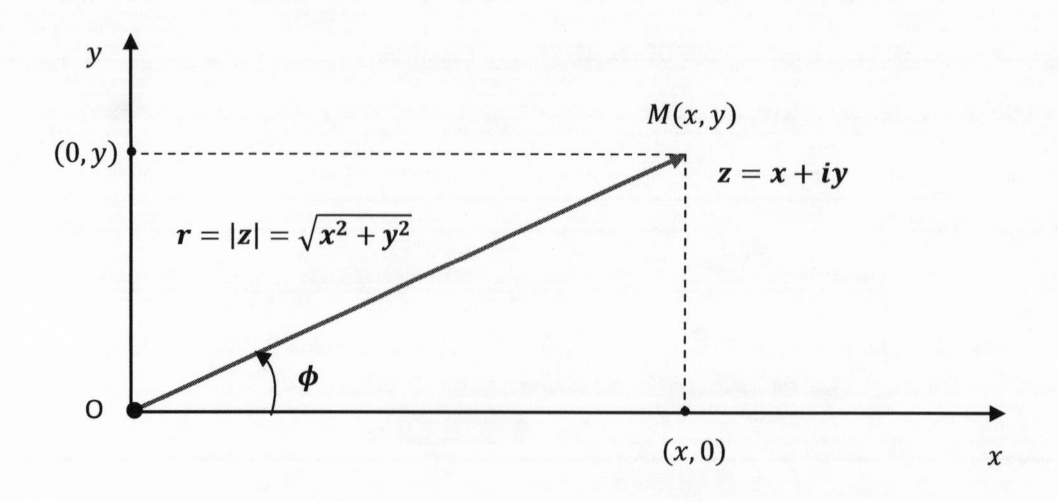

Fig.8-1: Cartesian and Polar form of a complex number z.

This plane is conveniently called, **the Complex Plane**. On this complex plane, the vector $\overrightarrow{OM}$, where $M(x, y)$ is the point (x, y) represents the complex number $x + iy$ as well. This is known as **the geometrical representation of the complex number $z = x + iy$.**

Another way of expressing a complex number $z = x + iy$, is the following:

From $z = x + iy \Longrightarrow z = \sqrt{x^2 + y^2}\left\{\frac{x}{\sqrt{x^2+y^2}} + i\frac{y}{\sqrt{x^2+y^2}}\right\}$.

However, from Fig. 8-1,

$$r = |z| = \sqrt{x^2 + y^2}\,,\ \cos\phi = \frac{x}{\sqrt{x^2+y^2}}\,, \qquad \sin\phi = \frac{y}{\sqrt{x^2+y^2}}\,,$$

Therefore the number $z = x + iy$ can be expressed in an equivalent form,

$$\boldsymbol{z = r(\cos\phi + i\sin\phi)} \text{ **where** } \boldsymbol{r = \sqrt{x^2 + y^2} \geq 0} \text{ **and** } \boldsymbol{\tan\phi = \frac{y}{x}} \quad \text{(8-10)}$$

This expression for z, is known as **the polar or the trigonometric form of the complex number** z.

The angle ϕ is called the argument or the amplitude or just the phase of z, and we write $\boldsymbol{\phi = Arg(z)}$. We note that **the argument of z has infinitely many different values, differing from each other by an integral multiple of 2π.**

There exists, however **a unique value of ϕ**, denoted by $\mathbf{arg}\,(\boldsymbol{z})$, such that

$\boldsymbol{-\pi < \phi = \mathbf{arg}\,(z) \leq \pi}$, which is called **the principal value of the argument.**

As an Example let us consider the complex numbers $z_1 = 1 + i$, $z_2 = -1 + i$ and $z_3 = 1 - i$, as shown in Fig. 8-2.

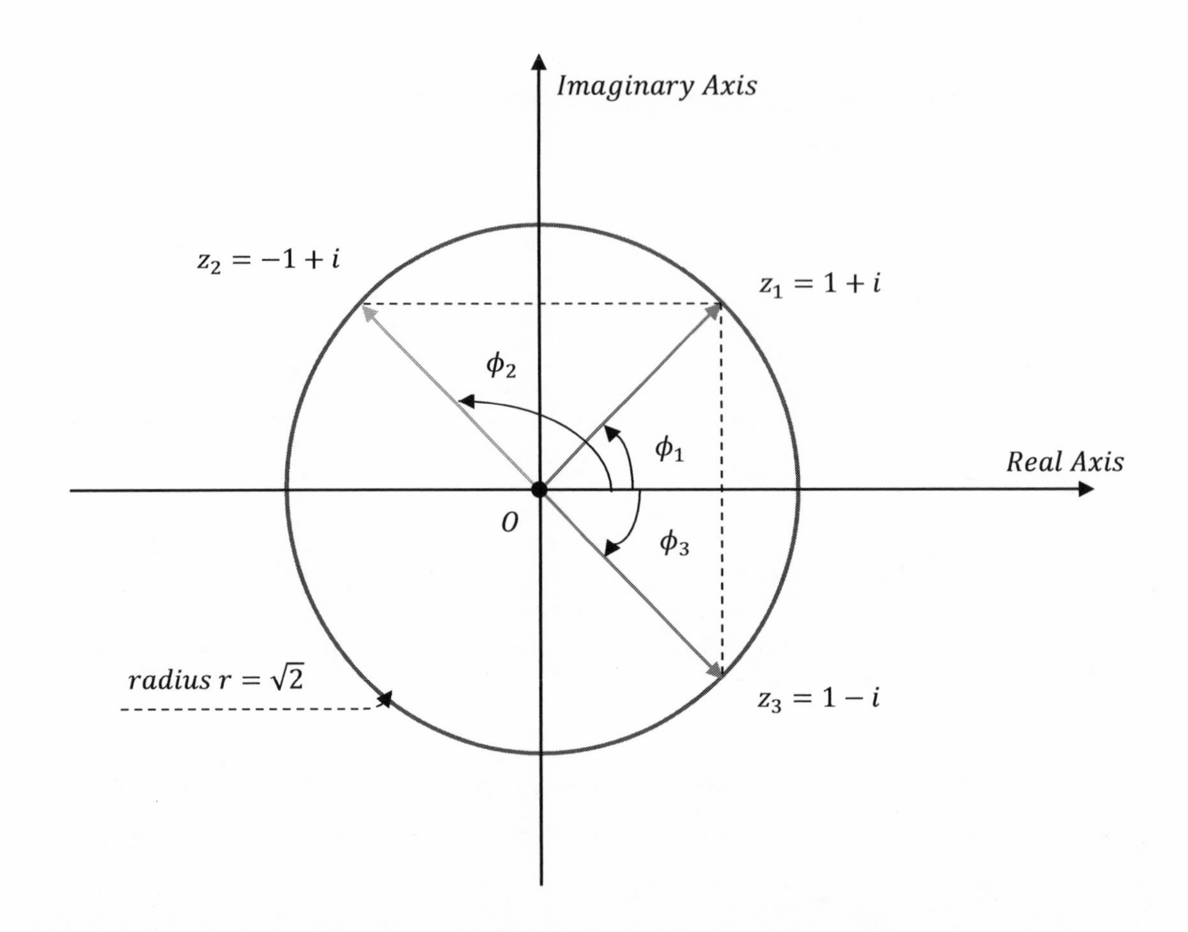

Fig. 8-2: Cartesian and Polar form of z_1, z_2, z_3.

The number z_1 has modulus $r_1 = |z_1| = \sqrt{1^2 + 1^2} = \sqrt{2}$, and phase $\phi_1 = \frac{\pi}{4}$, therefore $\boldsymbol{z_1 = \sqrt{2}\left(\cos\frac{\pi}{4} + i\sin\frac{\pi}{4}\right)}$.

The number z_2 has modulus $r_2 = |z_2| = \sqrt{(-1)^2 + 1^2} = \sqrt{2}$, and phase $\phi_2 = \frac{3\pi}{4}$, therefore $\boldsymbol{z_2 = \sqrt{2}\left(\cos\frac{3\pi}{4} + i\sin\frac{3\pi}{4}\right)}$.

The number z_3 has modulus $r_3 = |z_3| = \sqrt{(1)^2 + (-1)^2} = \sqrt{2}$, and phase $\phi_3 = -\frac{\pi}{4}$, therefore

$$\boldsymbol{z_3 = \sqrt{2}\left(\cos(-\tfrac{\pi}{4}) + i\sin(-\tfrac{\pi}{4})\right) = \sqrt{2}\left(\cos\tfrac{\pi}{4} - i\sin\tfrac{\pi}{4}\right)}.$$

Any **number** $\boldsymbol{x} > 0$ on the positive $x-$ semi axis can be written as $\boldsymbol{x}(\cos 0 + i \sin 0)$, while any number $\boldsymbol{x} < 0$ on the negative x −semi axis can be written as $|\boldsymbol{x}|(\cos \pi + i \sin \pi)$. For example the negative number $-3 = 3(\cos \pi + i \sin \pi)$.

On the other hand **points lying on the imaginary axis, will have a phase either** $\frac{\pi}{2}$ **or** $-\frac{\pi}{2}$. For example the imaginary unit in polar form can be written as $i = 1\left(\cos\frac{\pi}{2} + i \sin\frac{\pi}{2}\right)$, while

$$-i = 1\left(\cos\left(-\frac{\pi}{2}\right) + i \sin\left(-\frac{\pi}{2}\right)\right) = 1\left(\cos\frac{\pi}{2} - i \sin\frac{\pi}{2}\right).$$

(Recall that **the modulus of a complex number** $r = |z|$ **is always a positive number,** (see equation 8-10)).

The **arithmetic operations with complex numbers** are performed according to the well known rules of Algebra for operation on the binomials $(x + iy)$, after which i^2 is replaced by -1.

We note that,
$\boldsymbol{i^2 = -1,\ i^3 = i^2 \cdot i = -i,\ i^4 = i^2 \cdot i^2 = (-1)\cdot(-1) = 1,\ i^5 = i^4 \cdot i = i,}$
etc. In general $\boldsymbol{i^{4k+m} = i^m}$ **, where** $\boldsymbol{m = 0, 1, 2, 3.}$ (8-11)

The Cartesian form of z is more convenient for addition and subtraction, while the polar form is more convenient for multiplication and division.

The following two Theorems are of particular importance.

Theorem 8-1.

If $\boldsymbol{z_1 = r_1(\cos \phi_1 + i \sin \phi_1)}$ **and** $\boldsymbol{z_2 = r_2(\cos \phi_2 + i \sin \phi_2)}$**, then**

a) $\boldsymbol{z_1 z_2 = r_1 r_2\{\cos(\phi_1 + \phi_2) + i \sin(\phi_1 + \phi_2)\}}$ **and** (8-12)

b) $\boldsymbol{\frac{z_1}{z_2} = \frac{r_1}{r_2}\{\cos(\phi_1 - \phi_2) + i \sin(\phi_1 - \phi_2)\},\quad z_2 \neq 0 + i0.}$ (8-13)

For a proof see Example 8-6.

Note: Obviously (8-12) can be extended to any number of terms $z_k = r_k(\cos \phi_k + i \sin \phi_k),\ k = 1,2,3,\cdots,n,$ in which case (8-12) becomes, (using the Σ and Π notation),

$$\prod_{k=1}^{n} z_k = \{\prod_{k=1}^{n} r_k\} \cdot \{\cos(\sum_{k=1}^{n} \phi_k) + i\sin(\sum_{k=1}^{n} \phi_k)\}. \qquad (8\text{-}14)$$

In the case where $z_1 = z_2 = z_3 = \cdots = z_n = r(\cos\phi + i\sin\phi)$, equation (8-14) reduces to the following Theorem, known as **the De Moivre's Theorem**.

Theorem 8-2 (De Moivre's Theorem).

If n is any positive integer, then

$$(\cos\boldsymbol{\phi} + i\sin\boldsymbol{\phi})^n = \cos(n\boldsymbol{\phi}) + i\sin(n\boldsymbol{\phi}). \qquad (8\text{-}15)$$

Proof: The proof is easy and follows directly from eq.(8-14). Let the reader complete the proof.

De Moivre's formula is very important, and can be used **to find the values of trigonometric functions of multiple arcs**. As an example, for $n = 3$, (8-15) yields,

$(\cos\phi + i\sin\phi)^3 = \cos(3\phi) + i\sin(3\phi)$, or

$(\cos\phi)^3 + 3(\cos\phi)^2 (i\sin\phi) + 3\cos\phi\,(i\sin\phi)^2 + (i\sin\phi)^3 = \cos(3\phi) + i\sin(3\phi)$, or $(\cos\phi)^3 - 3\cos\phi\,(\sin\phi)^2 + i(3\sin\phi\,(\cos\phi)^2 - (\sin\phi)^3) = \cos(3\phi) + i\sin(3\phi)$, from which one finally obtains,

$\boldsymbol{\cos(3\phi) = (\cos\phi)^3 - 3\cos\phi\,(\sin\phi)^2}$ **and**
$\boldsymbol{\sin(3\phi) = 3\sin\phi\,(\cos\phi)^2 - (\sin\phi)^3}$.

B) The Hyperbolic Functions with Real Arguments.

In Example 5-2, the exponential function e^x $(-\infty < x < \infty)$ was defined by means of the infinite series,

$$e^x = \sum_{n=0}^{\infty} \frac{x^n}{n!} = 1 + \frac{x}{1!} + \frac{x^2}{2!} + \frac{x^3}{3!} + \frac{x^4}{4!} + \cdots, \qquad -\infty < x < \infty \quad (8\text{-}16)$$

Replacing x by $-x$, in (8-16), we obtain an expression for e^{-x}, i.e.

$$e^{-x} = \sum_{n=0}^{\infty} (-1)^n \frac{x^n}{n!} = 1 - \frac{x}{1!} + \frac{x^2}{2!} - \frac{x^3}{3!} + \frac{x^4}{4!} - \cdots \quad -\infty < x < \infty \quad (8\text{-}17)$$

In terms of e^x and e^{-x}, we may define the so called **Hyperbolic Functions of a real argument x**, by means of the following formulas:

i) The Hyperbolic cosine $\cosh x = \frac{e^x+e^{-x}}{2}$, (8-18)

ii) The Hyperbolic sine $\sinh x = \frac{e^x-e^{-x}}{2}$, (8-19)

iii) The Hyperbolic tangent $\tanh x = \frac{\sinh x}{\cosh x} = \frac{e^x-e^{-x}}{e^x+e^{-x}}$, (8-20)

iv) The Hyperbolic cotangent $\coth x = \frac{\cosh x}{\sinh x} = \frac{1}{\tanh x} = \frac{e^x+e^{-x}}{e^x-e^{-x}}$, (8-21)

The Hyperbolic Cosine, Sine and Tangent are defined $\forall \boldsymbol{x} \in \mathbb{R}$, while the Hyperbolic Cotangent is defined $\forall \boldsymbol{x} \in \mathbb{R} - \{\mathbf{0}\}$, due to the $\sinh x$ in the denominator in (8-21), which vanishes at $x = 0$.
Making use of (8-16),(8-17),(8-18) and (8-19), one may find the following **series representation** for the $\cosh x$ and $\sinh x$ functions.

$$\cosh x = 1 + \frac{x^2}{2!} + \frac{x^4}{4!} + \frac{x^6}{6!} + \frac{x^8}{8!} + \cdots, \qquad -\infty < x < \infty \qquad (8\text{-}22)$$

$$\sinh x = x + \frac{x^3}{3!} + \frac{x^5}{5!} + \frac{x^7}{7!} + \frac{x^9}{9!} + \cdots, \qquad -\infty < x < \infty \qquad (8\text{-}23)$$

A rough sketch of the functions $\cosh x$, $\sinh x$ and $\tanh x$ is shown in Fig. 8-3.

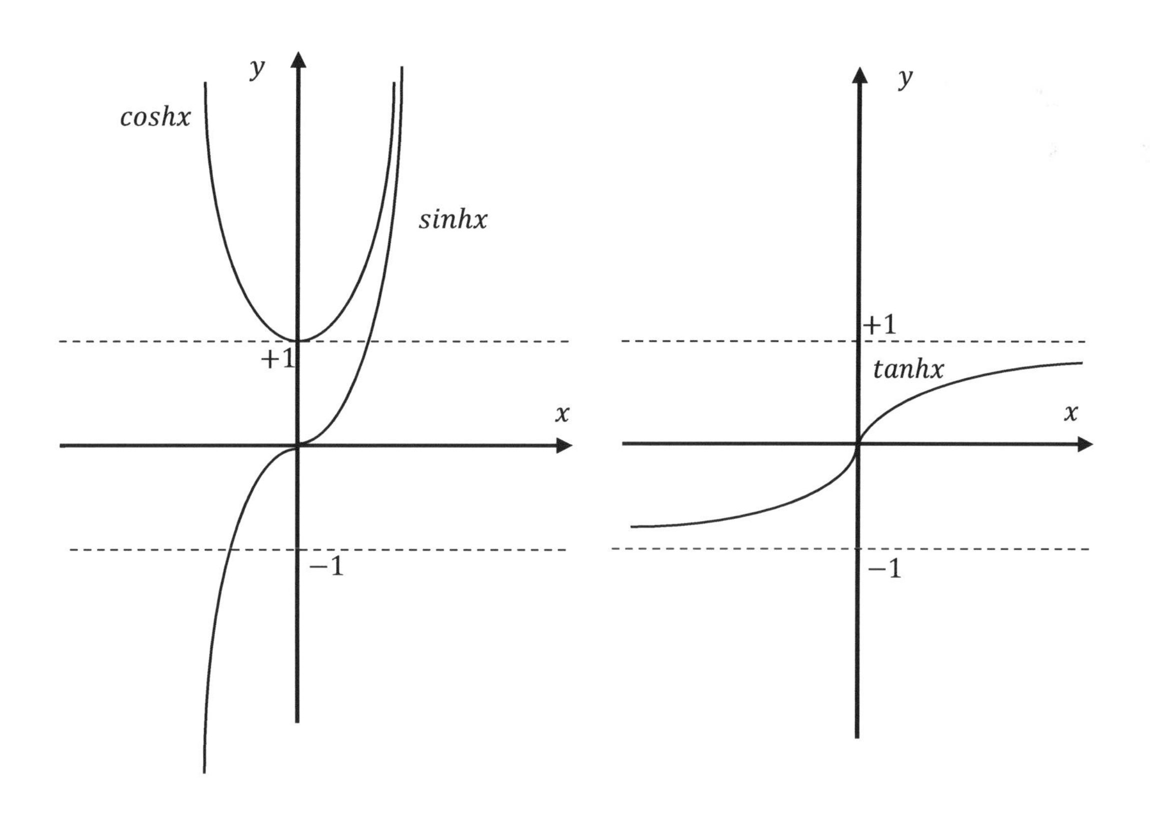

Fig. 8-3: Graph of the functions $\cosh x$, $\sinh x$ and $\tanh x$.

It is not difficult to show that $\boldsymbol{\cosh x}$ **is an even function of** $\boldsymbol{x}$, i.e. $\boldsymbol{\cosh(-x) = \cosh x}$, while $\boldsymbol{\sinh x}$ **is an odd function of** $\boldsymbol{x}$**, i.e.** $\boldsymbol{\sinh(-x) = -\sinh x}$.

Also $\lim_{x\to+\infty} \cosh x = +\infty$, $\lim_{x\to+\infty} \sinh x = +\infty$, $\lim_{x\to-\infty} \cosh x = +\infty$, and $\lim_{x\to-\infty} \sinh x = -\infty$.

The function $\boldsymbol{\tanh x}$ **is an odd function of** $\boldsymbol{x}$**, i.e.** $\boldsymbol{\tanh(-x) = -\tanh x}$**,** while $\lim_{x\to+\infty} \tanh x = +1$, and $\lim_{x\to-\infty} \tanh x = -1$.

Finally we remind the reader, that the well known Trigonometric Functions $\cos x$ and $\sin x$ admit the following series representation,

$$\boldsymbol{\cos x = 1 - \frac{x^2}{2!} + \frac{x^4}{4!} - \frac{x^6}{6} + \frac{x^8}{8!} - \frac{x^{10}}{10!} + \cdots}, \qquad -\infty < x < \infty \qquad \text{(8-24)}$$

$$\boldsymbol{\sin x = x - \frac{x^3}{3!} + \frac{x^5}{5!} - \frac{x^7}{7!} + \frac{x^9}{9!} - \frac{x^{11}}{11!} + \cdots}, \qquad -\infty < x < \infty \qquad \text{(8-25)}$$

C) The Exponential Function e^z, $z \in \mathbb{C}$.

Let $z = x + iy$ be any complex number.The **exponential function** $\boldsymbol{e^z}$ is defined as

$$\boldsymbol{e^z} = \sum_{n=0}^{\infty} \frac{z^n}{n!} = 1 + \frac{z}{1!} + \frac{z^2}{2!} + \frac{z^3}{3!} + \frac{z^4}{4!} + \frac{z^5}{5!} + \cdots, \quad z \in \mathbb{C}. \qquad \text{(8-26)}$$

In the next chapter, we shall prove that **the infinite series in (8-26), converges for any complex number** $\boldsymbol{z}$**, and the sum of this series is by definition the exponential function** $\boldsymbol{e^z}$**.** If $z = x + i0 = x \in \mathbb{R}$, (8-26) reduces to (8-16), i.e. to the series representation of e^x. If z is a pure imaginary number $\boldsymbol{z = 0 + i\phi = i\phi}$, ($\boldsymbol{\phi} \in \mathbb{R}$), (8-26) implies,

$$e^{i\phi} = 1 + \frac{i\phi}{1!} + \frac{(i\phi)^2}{2!} + \frac{(i\phi)^3}{3!} + \frac{(i\phi)^4}{4!} + \frac{(i\phi)^5}{5!} + \frac{(i\phi)^6}{6!} + \frac{(i\phi)^7}{7!} + \cdots, \text{ or}$$

$$e^{i\phi} = \left\{1 - \frac{\phi^2}{2!} + \frac{\phi^4}{4!} - \frac{\phi^6}{6!} + \cdots\right\} + i\left\{\phi - \frac{\phi^3}{3!} + \frac{\phi^5}{5!} - \frac{\phi^7}{7!} + \cdots\right\},$$

Or finally, making use of (8-24) and (8-25) we have,

$$\boldsymbol{e^{i\phi} = \cos\phi + i\sin\phi}. \qquad \text{(8-27)}$$

Replacing ϕ by $(-\phi)$ and noting that $\cos(-\boldsymbol{\phi}) = \cos\boldsymbol{\phi}$ and that $\sin(-\boldsymbol{\phi}) = -\sin\boldsymbol{\phi}$, equation (8-27) becomes,

$$e^{-i\phi} = \cos\phi - i\sin\phi. \qquad (8\text{-}28)$$

From (8-25) and (8-26), one easily obtains,

$$\cos\phi = \frac{e^{i\phi}+e^{-i\phi}}{2} \quad \text{and} \quad \sin\phi = \frac{e^{i\phi}-e^{-i\phi}}{2i}. \qquad (8\text{-}29)$$

Formulas (8-27) (8-28) and (8-29) are **the famous Euler's formulas**.

At $\phi = \pi$, equation (8-27) yields,

$$e^{i\pi} + 1 = 0 \qquad (8\text{-}30)$$

one of the most noted formulas in Mathematics, since it relates in a simple and elegant way the five mathematical constants $\{0, 1, e, \pi, i\}$.

D) Trigonometric and Hyperbolic Functions with Complex Arguments.

Making use of the Euler's formulas one may **extend the definition of the Trigonometric and the Hyperbolic Functions, from real to complex arguments**.

Let $z = x + iy$ be any complex number. We define:

i) $\cos z = \dfrac{e^{iz}+e^{-iz}}{2}$ $\qquad$ $\cosh z = \dfrac{e^{z}+e^{-z}}{2}$ $\qquad$ (8-31)

ii) $\sin z = \dfrac{e^{iz}-e^{-iz}}{2i}$ $\qquad$ $\sinh z = \dfrac{e^{z}-e^{-z}}{2}$ $\qquad$ (8-32)

iii) $\tan z = \dfrac{\sin z}{\cos z}$ $\qquad$ $\tanh z = \dfrac{\sinh z}{\cosh z}$ $\qquad$ (8-33)

If $z = x + i0 = x \in \mathbb{R}$, (8-31), (8-32) and (8-33) reduce back to the familiar definitions of the corresponding functions with real arguments.

The Trigonometric and Hyperbolic functions with complex arguments, **share many common properties with the corresponding functions of real arguments**. For example it can be shown that,

$$(\cos z)^2 + (\sin z)^2 = 1$$

$$\sin(z_1 \pm z_2) = \sin z_1 \cos z_2 \pm \sin z_2 \cos z_1$$

$$\cos(z_1 \pm z_2) = \cos z_1 \cos z_2 \mp \sin z_1 \sin z_2$$

$$\tan(z_1 \pm z_2) = \frac{\tan z_1 \pm \tan z_2}{1 \mp \tan z_1 \tan z_2},$$

$$(\cosh z)^2 - (\sinh z)^2 = 1$$

$$\sinh(z_1 \pm z_2) = \sinh z_1 \cosh z_2 \pm \sinh z_2 \cosh z_1$$

$$\cosh(z_1 \pm z_2) = \cosh z_1 \cosh z_2 \pm \sinh z_1 \sinh z_2$$

$$\tanh(z_1 \pm z_2) = \frac{\tanh z_1 \pm \tanh z_2}{1 \pm \tanh z_1 \tanh z_2}.$$

However, **there are also some striking differences** between them, the most characteristic one being that while **the functions $\sin x$ and $\cos x$ are bounded, ($|\sin x| \leq 1$ and $|\cos x| \leq 1$), the $|\sin z|$ and $|\cos z|$ are unbounded, ($\lim_{|z|\to\infty}|\sin z| = \infty$ and $\lim_{|z|\to\infty}|\cos z| = \infty$),** (for a proof see Problem 8-4).

If **$z = i\phi$ ($\phi \in \mathbb{R}$),** is a pure imaginary number, one can easily show that,

$$\cos(\phi i) = \cosh\phi \quad \text{and} \quad \cos\phi = \cosh(i\phi) \tag{8-34}$$

$$\sin(i\phi) = i\sinh\phi \quad \text{and} \quad \sin\phi = -i\sinh(i\phi) \tag{8-35}$$

For a proof of (8-34) and (8-35) see Example 8-7.

Example 8-1.

If $z_1 = 1 + 2i,\ z_2 = 3 + i$, find $z_1 + z_2,\ z_1z_2$ and $z_1{}^2z_2$.

Solution

$z_1 + z_2 = (1 + 2i) + (3 + i) = (1 + 3) + i(2 + 1) = 4 + 3i.$

$z_1z_2 = (1 + 2i)(3 + i) = 3 + 6i + i + 2i^2 = 3 - 2 + 7i = 1 + 7i.$

$z_1{}^2z_2 = (1 + 2i)^2(3 + i) = (1^2 + 2 \cdot 1 \cdot (2i) + (2i)^2)(3 + i) = (1 + 4i - 4)(3 + i) = (-3 + 4i)(3 + i) = -9 + 12i - 3i + 4i^2 = -9 - 4 + 9i = -13 + 9i.$

Example 8-2.

If $z_1 = 2 + 3i$ and $z_2 = 4 + 5i$, find $\frac{z_1}{z_2}$.

Solution

Let $w = \frac{z_1}{z_2} = \frac{2+3i}{4+5i}$.

We multiply both numerator and denominator by the complex conjugate of the denominator, to get

$$w = \frac{2+3i}{4+5i} = \frac{(2+3i)(4-5i)}{(4+5i)(4-5i)} = \frac{8+12i-10i-15i^2}{4^2+5^2} = \frac{8+15+2i}{41} = \frac{23+2i}{41} = \frac{23}{41} + \frac{2}{41}i.$$

Note: In general, $\frac{z_1}{z_2} = \frac{z_1\overline{z_2}}{z_2\overline{z_2}} = \frac{z_1\overline{z_2}}{|z_2|^2}$ (see (8-4)).

Example 8-3.

If z and w are any two complex numbers, show that

$$z\bar{z} = |z|^2, \quad \overline{z+w} = \bar{z} + \bar{w}, \quad \overline{z \cdot w} = \bar{z} \cdot \bar{w}, \quad |z + w| \leq |z| + |w|.$$

(Obviously these properties can be generalized **to any number of terms,** see (8-5), (8-6), (8-7) and (8-8)).

Solution

Let $z = x + iy$ and $w = a + ib$. Then,

i) $z\bar{z} = (x + iy)(x - iy) = x^2 + iyx - ixy - i^2y^2 = x^2 + y^2 =$
$= \left(\sqrt{x^2 + y^2}\right)^2 = |z|^2$

ii) $\overline{z + w} = \overline{(x + iy) + (a + ib)} = \overline{(x + a) + i(y + b)} = (x + a) -$
$i(y + b) = \quad (x - iy) + (a - ib) = \bar{z} + \bar{w}.$

iii) $\overline{zw} = \overline{(x + iy)(a + ib)} = \overline{ax + iay + ibx - by} =$
$\overline{(ax - by) + i(ay + bx)} = \ (ax - by) - i(ay + bx) =$
$= (x - iy)(a - ib) = \bar{z}\bar{w}.$

iv) $|z+w| = |(x+iy)+(a+ib)| = |(x+a)+i(y+b)| = \sqrt{(x+a)^2+(y+b)^2}$.

We want to show that

$$|z+w| \leq |z|+|w| \Leftrightarrow \sqrt{(x+a)^2+(y+b)^2} \leq \sqrt{x^2+y^2}+\sqrt{a^2+b^2}. \qquad (*)$$

Since **both sides of (*) are non negative**, it suffices to show that

$$\left\{\sqrt{(x+a)^2+(y+b)^2}\right\}^2 \leq \left\{\sqrt{x^2+y^2}+\sqrt{a^2+b^2}\right\}^2$$

$$\Leftrightarrow$$

$$x^2+a^2+2ax+y^2+b^2+2by \leq x^2+y^2+a^2+b^2+2\sqrt{x^2+y^2}\sqrt{a^2+b^2} \Leftrightarrow$$

$$ax+by \leq \sqrt{x^2+y^2}\sqrt{a^2+b^2}\,. \qquad (**)$$

If $(\boldsymbol{ax}+\boldsymbol{by}) < 0$, (**) is obviously true.

If $(\boldsymbol{ax}+\boldsymbol{by}) \geq \mathbf{0}$, in order to show (**), it suffices to show that

$$(ax+by)^2 \leq \left\{\sqrt{x^2+y^2}\sqrt{a^2+b^2}\right\}^2 \Leftrightarrow$$

$$a^2x^2+b^2y^2+2abxy \leq a^2x^2+a^2y^2+b^2x^2+b^2y^2 \Leftrightarrow$$

$$2abxy \leq a^2y^2+b^2x^2 \Leftrightarrow (ay-bx)^2 \geq 0$$

which obviously is true, and this completes the proof.

Note: We could have proved that $|z+w| \leq |z|+|w|$ considering **the geometrical representation of the complex numbers $\boldsymbol{z}$ and $\boldsymbol{w}$** as shown in Fig. 8-4.

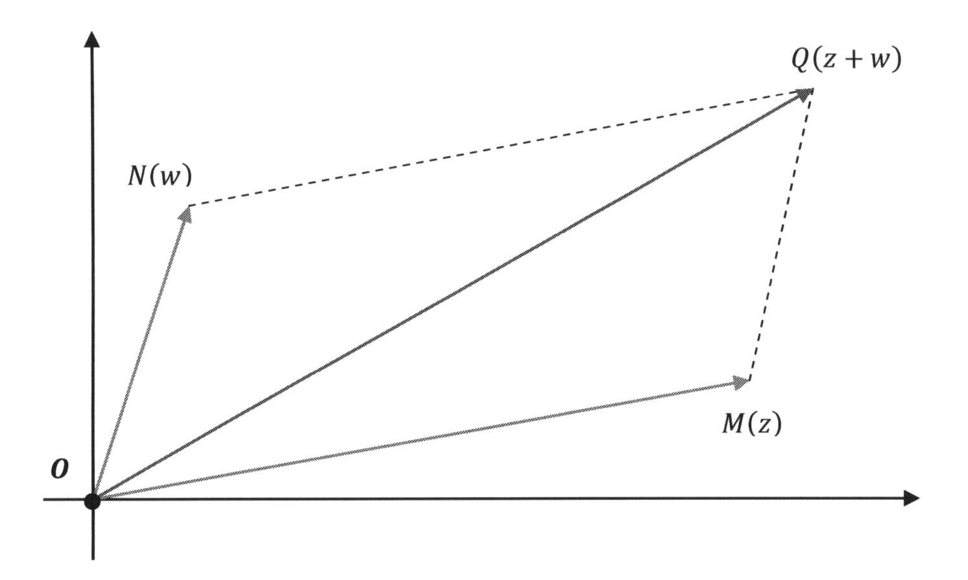

Fig. 8-4: Geometrical addition of two complex numbers z and w.

Let $\overrightarrow{OM}$ be the vector corresponding to z and $\overrightarrow{ON}$ the vector corresponding to w. The vector sum $\overrightarrow{OQ} = \overrightarrow{OM} + \overrightarrow{ON}$ will thus represent the complex number $(z+w)$. From the triangle OMQ we have, $|\overrightarrow{OQ}| \leq |\overrightarrow{OM}| + |\overrightarrow{MQ}|$ or $|\overrightarrow{OQ}| \leq |\overrightarrow{OM}| + |\overrightarrow{ON}|$, since $\overrightarrow{ON} = \overrightarrow{MQ}$ and therefore $|\overrightarrow{ON}| = |\overrightarrow{MQ}|$, or finally $|z+w| \leq |z| + |w|$, and the proof is completed.

Example 8-4.

If z and w are two complex numbers, show that $|z \cdot w| = |z| \cdot |w|$, and that $\left|\frac{z}{w}\right| = \frac{|z|}{|w|}$, provided $|w| \neq 0$.

Solution

i) Let $z = x + iy$ and $w = a + ib$. Then $|z| = \sqrt{x^2 + y^2}$ and $|w| = \sqrt{a^2 + b^2}$ therefore $|z \cdot w| = |(x+iy)(a+ib)| = |(ax - by) + i(ay + bx)| = \sqrt{(ax - by)^2 + (ay + bx)^2} = \sqrt{a^2x^2 + b^2y^2 + a^2y^2 + b^2x^2} = \sqrt{(x^2 + y^2)(a^2 + b^2)} = \sqrt{x^2 + y^2}\sqrt{a^2 + b^2} = |z| \cdot |w|$, and the proof is completed.

ii) If we set $s = \frac{z}{w}$ then $z = s \cdot w$, so according to **(i)**, $|z| = |s \cdot w| = |s| \cdot |w|$ or $\frac{|z|}{|w|} = |s| = \left|\frac{z}{w}\right|$, and this completes the proof.

Example 8-5.

Express the complex number $z = 1 + i\sqrt{3}$ in polar form, and then find the number $w = z^{103}$.

Solution

i) The modulus of z is $|z| = \sqrt{1^2 + \left(\sqrt{3}\right)^2} = \sqrt{4} = 2$, while **its principal argument $\boldsymbol{\phi}$** satisfies $\tan\phi = \frac{\sqrt{3}}{1}$ or equivalently, $\phi = \frac{\pi}{3}$, and $\boldsymbol{z} = \mathbf{2}\left(\mathbf{cos}\frac{\pi}{3} + \boldsymbol{i}\,\mathbf{sin}\frac{\pi}{3}\right)$.

ii) Making use of the **De Moivre's Theorem,**

$$w = z^{103} = \left\{2\left(\cos\frac{\pi}{3} + i\sin\frac{\pi}{3}\right)\right\}^{103} = 2^{103}\left\{\cos\left(\frac{103 \cdot \pi}{3}\right) + i\sin\left(\frac{103 \cdot \pi}{3}\right)\right\} =$$

$$2^{103}\left\{\cos\left(2 \cdot 17 \cdot \pi + \frac{\pi}{3}\right) + i\sin\left(2 \cdot 17 \cdot \pi + \frac{\pi}{3}\right)\right\} =$$

$$= 2^{103}\left(\cos\frac{\pi}{3} + i\sin\frac{\pi}{3}\right) = 2^{103}\left(\frac{1}{2} + i\frac{\sqrt{3}}{2}\right) = 2^{102}\left(1 + i\sqrt{3}\right).$$

Example 8-6.

Prove Theorem 8-1.

Solution

i) The product

$$\begin{aligned} z_1 z_2 = \{r_1(\cos\phi_1 + i\sin\phi_1)\}\{r_2(\cos\phi_2 + i\sin\phi_2)\} \\ = r_1 r_2\{(\cos\phi_1\cos\phi_2 - \sin\phi_1\sin\phi_2) \\ + i(\sin\phi_1\cos\phi_2 + \sin\phi_2\cos\phi_1)\} \\ = r_1 r_2\{\cos(\phi_1 + \phi_2) + i\sin(\phi_1 + \phi_2)\} \end{aligned}$$

and this completes the proof.

ii) The ratio

$$\frac{z_1}{z_2}=\frac{r_1(\cos\phi_1+i\sin\phi_1)}{r_2(\cos\phi_2+i\sin\phi_2)}=\frac{r_1}{r_2}\cdot\frac{(\cos\phi_1+i\sin\phi_1)(\cos\phi_2-i\sin\phi_2)}{|\cos\phi_2+i\sin\phi_2|^2}=$$

$$\frac{r_1}{r_2}\cdot\{(\cos\phi_1\cos\phi_2+\sin\phi_1\sin\phi_2)+i(\sin\phi_1\cos\phi_2-\sin\phi_2\cos\phi_1)\}=$$

$$\frac{r_1}{r_2}\cdot\{\cos(\phi_1-\phi_2)+i\sin(\phi_1-\phi_2)\}.$$

and the proof is completed.

(We note that $|\boldsymbol{\cos\phi_2+i\sin\phi_2}|^2=(\boldsymbol{\cos\phi_2})^2+(\boldsymbol{\sin\phi_2})^2=\mathbf{1}$).

Example 8-7.

Prove formulas (8-34) and (8-35).

Solution

Making use of **the Euler's formulas**, we have,

i) $\cos(i\phi)=\frac{e^{i(i\phi)}+e^{-i(i\phi)}}{2}=\frac{e^{-\phi}+e^{\phi}}{2}=\cosh\phi.$

ii) $\cos\phi=\frac{e^{i\phi}+e^{-i\phi}}{2}=\cosh(i\phi)$, (see equation (8-31)).

iii) $\sin(i\phi)=\frac{e^{i(i\phi)}-e^{-i(i\phi)}}{2i}=\frac{e^{-\phi}-e^{\phi}}{2i}=\left(-\frac{1}{i}\right)\frac{e^{\phi}-e^{-\phi}}{2}=i\sinh\phi.$

iv) $\sin\phi=\frac{e^{i\phi}-e^{-i\phi}}{2i}=\left(\frac{1}{i}\right)\frac{e^{i\phi}-e^{-i\phi}}{2}=-i\sinh(i\phi).$

Example 8-8.

i) Express $\cos(5\phi)$ and $\sin(5\phi)$ in terms of $\cos\phi$ and $\sin\phi$.

ii) Express $\tan(5\phi)$ in terms of $\tan\phi$.

Solution

i) Making use of **the De Moivre's Theorem**, we have,

$(\cos\phi + i\sin\phi)^5 = \cos(5\phi) + i\sin(5\phi)$, and expanding the left side term, making use of **the Newton's Binomial Formula**, we obtain,

$\sum_{k=0}^{5}\binom{5}{k}(\cos\phi)^{5-k}(i\sin\phi)^k = \cos(5\phi) + i\sin(5\phi)$ or

$$\binom{5}{0}(\cos\phi)^5 + \binom{5}{1}(\cos\phi)^4(i\sin\phi) + \binom{5}{2}(\cos\phi)^3(i\sin\phi)^2 + \binom{5}{3}(\cos\phi)^2(i\sin\phi)^3 + \binom{5}{4}\cos\phi\,(i\sin\phi)^4 + \binom{5}{5}(i\sin\phi)^5 = \cos(5\phi) + i\sin(5\phi),$$

and equating real and imaginary parts, we have,

$$\mathbf{\cos(5\phi) = (\cos\phi)^5 - 10(\cos\phi)^3(\sin\phi)^2 + 5\cos\phi\,(\sin\phi)^4}, \text{ and } (*)$$

$$\mathbf{\sin(5\phi) = (\sin\phi)^5 - 10(\cos\phi)^2(\sin\phi)^3 + 5\sin\phi\,(\cos\phi)^4}. \qquad (**)$$

ii) The $\mathbf{\tan(5\phi)} = \frac{\mathbf{\sin(5\phi)}}{\mathbf{\cos(5\phi)}} = \frac{\mathbf{(\tan\phi)^5 - 10(\tan\phi)^3 + 5\tan\phi}}{\mathbf{1 - 10(\tan\phi)^2 + 5(\tan\phi)^4}}$.

The last expression is obtained from (*) and (**), since

$\tan(5\phi) = \frac{\sin(5\phi)}{\cos(5\phi)} = \frac{\sin(5\phi)\div(\cos\phi)^5}{\cos(5\phi)\div(\cos\phi)^5}$.

Example 8-9.

Let $A_1A_2A_3\cdots A_n$ be a regular polygon with n sides, inscribed in a circle of radius R. Evaluate, in terms of R, the product $(A_1A_2)(A_1A_3)(A_1A_4)\cdots(A_1A_{n-1})(A_1A_n)$.

Solution

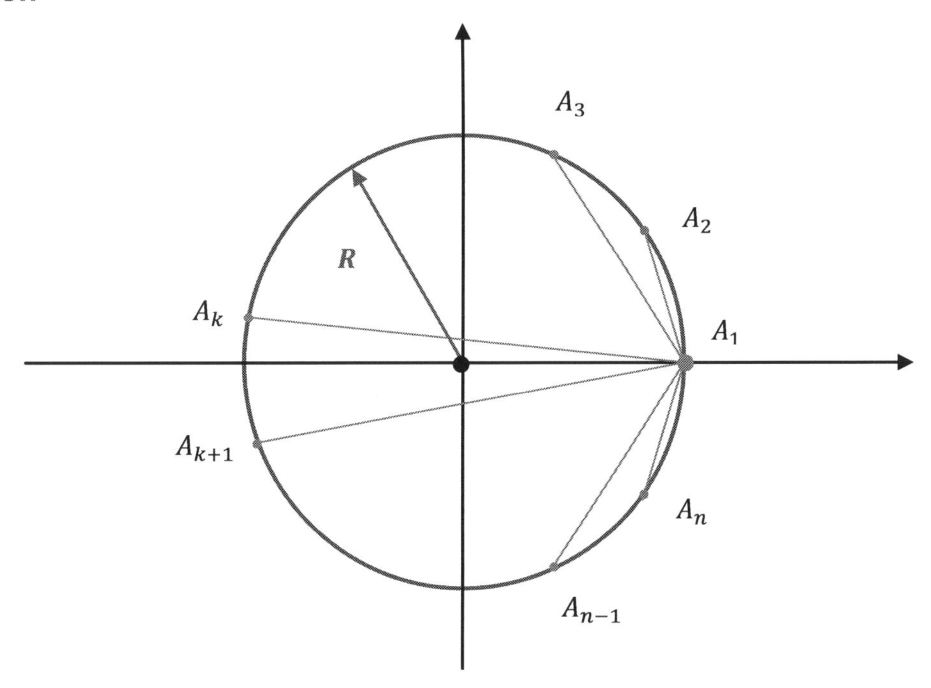

Fig.8-5: Regular polygon with n sides.

The points A_{k+1} $(k = 0,1,2,3,\cdots,n-1)$ on the complex plane, represent the complex numbers $z_k = R\left\{\cos\left(k\frac{2\pi}{n}\right) + i\sin\left(k\frac{2\pi}{n}\right)\right\}$ where $k = 0,1,2,\cdots(n-1)$.

Since $\boldsymbol{z_k}^{\boldsymbol{n}} = \boldsymbol{R}^{\boldsymbol{n}}$, as verified easily by **the De Moivre's Theorem,** we conclude **that z_k are the roots of the polynomial equation $\boldsymbol{z^n - R^n = 0}$**, and therefore,

$z^n - R^n = (z - z_0)(z - z_1)(z - z_2)\cdots(z - z_{n-1})$, or since $z_0 = R$,

$\frac{z^n - R^n}{z - R} = (z - z_1)(z - z_2)\cdots(z - z_{n-1})$, or even more,

$$\boldsymbol{z^{n-1} + Rz^{n-2} + R^2z^{n-3} + \cdots + R^{n-2}z + R^{n-1} =}$$

$$\boldsymbol{= (z - z_1)(z - z_2)\cdots(z - z_{n-1}).} \qquad (*)$$

Equation (*) is an identity, true for all values of z, therefore it must be true for $z = R$ as well, in which case we have,

$$nR^n = (R - z_1)(R - z_2)\cdots(R - z_{n-1}). \qquad (**)$$

We note that **the length** (A_1A_2) **is the modulus of the complex number** $\boldsymbol{z_1 - z_0 = z_1 - R}$, i.e. $(\boldsymbol{A_1A_2}) = |\boldsymbol{z_1 - R}|$, and similarly, $(\boldsymbol{A_1A_3}) = |\boldsymbol{z_2 - R}|$, $(\boldsymbol{A_1A_4}) = |\boldsymbol{z_3 - R}|$, etc.

Taking the modulus of both sides in (**), we have,

$$|nR^{n-1}| = |(R - z_1)(R - z_2)\cdots(R - z_{n-1})| = |z_1 - R||z_2 - R|\cdots|z_{n-1} - R|,$$

and finally,

$$(\boldsymbol{A_1A_2})(\boldsymbol{A_1A_3})(\boldsymbol{A_1A_4})\cdots(\boldsymbol{A_1A_{n-1}})(\boldsymbol{A_1A_n}) = \boldsymbol{nR^{n-1}}.$$

Example 8-10.

If z and w are any two complex numbers show that:

i) $\sin(z + w) = \sin z \cos w + \sin w \cos z,$

ii) $(\sin z)^2 + (\cos z)^2 = 1,$

iii) $(\cosh z)^2 - (\sinh z)^2 = 1,$

iv) $\sinh z + \sinh w = 2\sinh(\frac{z+w}{2})\cosh(\frac{z-w}{2}).$

Solution

i) Making use of Euler's formulas we have,

$$\sin z \cos w + \sin w \cos z = \frac{e^{iz}-e^{-iz}}{2i} \cdot \frac{e^{iw}+e^{-iw}}{2} + \frac{e^{iw}-e^{-iw}}{2i} \cdot \frac{e^{iz}+e^{-iz}}{2} = \frac{e^{i(z+w)}-e^{-i(z+w)}}{2i} = \sin(z + w).$$

ii) $(\sin z)^2 + (\cos z)^2 = \left(\frac{e^{iz}-e^{-iz}}{2i}\right)^2 + \left(\frac{e^{iz}+e^{-iz}}{2}\right)^2 = \frac{4}{4} = 1.$

iii) $(\cosh z)^2 - (\sinh z)^2 = \left(\frac{e^{z}+e^{-z}}{2}\right)^2 - \left(\frac{e^{z}-e^{-z}}{2}\right)^2 = \frac{4}{4} = 1.$

iv) $2\sinh(\frac{z+w}{2})\cosh(\frac{z-w}{2}) = 2 \cdot \frac{e^{\left(\frac{z+w}{2}\right)}-e^{-\left(\frac{z+w}{2}\right)}}{2} \cdot \frac{e^{\left(\frac{z-w}{2}\right)}+e^{-\left(\frac{z-w}{2}\right)}}{2} = \frac{e^{z}-e^{-z}}{2} + \frac{e^{w}-e^{-w}}{2} = \sinh z + \sinh w.$

PROBLEMS

8-1) If $z = 1 - i$, express z in polar form, and then find z^{12}.

(**Answer:** $z = \sqrt{2}\left(\cos\frac{\pi}{4} - i\sin\frac{\pi}{4}\right)$, $z^{12} = -64$).

8-2) If $z = 2 + 3i$, $w = 1 + i$, find

z^2w^3, $\frac{z^2}{w^3}$, zw^6, $|zw|$, $\left|\frac{z}{w}\right|$.

8-3) Find $\left(1 + i\sqrt{3}\right)^5 + \left(1 - i\sqrt{3}\right)^5$ and $\left(1 + i\sqrt{3}\right)^5 \cdot \left(1 - i\sqrt{3}\right)^5$.

(**Answer:** 32, 4^5).

8-4) If x and y are real numbers, show that $\sin(x + iy) = \sin x \cosh y + i\cos x \sinh y$. Then show that
$|\mathbf{\sin z}| = |\mathbf{\sin(x + iy)}| = (\mathbf{\cosh y})^2 - (\mathbf{\cos x})^2 = (\mathbf{\sin x})^2 + (\mathbf{\sinh y})^2$.
Is $|\sin z|$ bounded as $|z| = \sqrt{x^2 + y^2} \to \infty$?

(Note that $\lim_{y\to\infty} \cosh y = \infty$ and $\lim_{y\to\infty} \sinh y = \infty$).

8-5) Find the roots of the equation $x^2 - 2x\cos\theta + 1 = 0$.

(**Answer:** $x_1 = \cos\theta + i\sin\theta$, $x_2 = \cos\theta - i\sin\theta$).

8-6) If z and w are complex numbers, show that
$\cos(z \pm w) = \cos z \cos w \mp \sin z \sin w$.

8-7) If n is any positive integer, find $(\cos\theta + i\sin\theta)^n(\sin\theta + i\cos\theta)^n$.

(**Answer:** $\cos\left(\frac{n\pi}{2}\right) + i\sin\left(\frac{n\pi}{2}\right) = e^{i\frac{n\pi}{2}}$).

8-8) Express $\sin(7\theta)$ and $\cos(7\theta)$ in terms of $\cos\theta$ and $\sin\theta$, and then express $\tan(7\theta)$ in terms of $\tan\theta$, (see Example 8-8).

8-9) Let z_1, z_2 and z_3 be three complex numbers.
If ${z_1}^2 + {z_2}^2 + {z_3}^2 = z_1z_2 + z_2z_3 + z_3z_1$, show that the points $A_1(z_1)$, $A_2(z_2)$ and $A_3(z_3)$ on the complex plane , are the vertices of **an equilateral triangle**.

Hint: It suffices to show that $|z_1 - z_2| = |z_2 - z_3| = |z_3 - z_1|$.

From the given conditions we have, $(z_1 - z_2)^2 + (z_2 - z_3)^2 + (z_3 - z_1)^2 = 0$, (why?). Let a, b and c be the three complex numbers, $a = z_1 - z_2$, $b = z_2 - z_3$, and $c = z_3 - z_1$. Then $\boldsymbol{a + b + c = 0}$ and $\boldsymbol{a^2 + b^2 + c^2 = 0}$, or equivalently,

$\frac{a}{c} + \frac{b}{c} = -1$ and $\left(\frac{a}{c}\right)^2 + \left(\frac{b}{c}\right)^2 = -1 \Rightarrow \frac{a}{c} \cdot \frac{b}{c} = 1$, i.e. the numbers $\frac{a}{c}$ and $\frac{b}{c}$ will be the roots of the quadratic equation $x^2 + x + 1 = 0$, etc.

8-10) If z and w are complex numbers, show that

$|z| - |w| \leq \big||z| - |w|\big| \leq |z \pm w| \leq |z| + |w|$.

8-11) If $\left|\frac{z+5}{z+1}\right| = \sqrt{5}$ show that $|z| = \sqrt{5}$.

Hint: For any complex number w, $w\bar{w} = |w|^2$.

8-12) Find the complex number $w = \left(\frac{1+i}{\sqrt{3}}\right)^{57} \left(\frac{1+i\sqrt{3}}{2}\right)^{82}$, and also find $|w|$.

Hint: Express the numbers in polar form and make use of **the De Moivre's Theorem**.

8-13) Solve for $x \in \mathbb{R}$ the equation $\cosh x = 3$.

Hint: $\cosh x = \frac{e^x + e^{-x}}{2} = 3 \Rightarrow e^{2x} - 6e^x + 1 = 0$, set $t = e^x$ etc.

(**Answer:** $x = \pm\ln\left(3 + \sqrt{8}\right)$).

8-14) If z is a complex number, $(z \neq 1)$, show that $1 + z + z^2 + z^3 + \cdots + z^n = \frac{1-z^{n+1}}{1-z}$. Then set $z = \cos\theta + i\sin\theta$ to obtain **a closed form expression** for $\sum_{k=0}^{n} \cos(k\theta)$ and $\sum_{k=0}^{n} \sin(k\theta)$.

8-15) Show that $\boldsymbol{\sin\frac{\pi}{n}\sin\frac{2\pi}{n}\sin\frac{3}{n}\cdots\sin\frac{(n-1)\pi}{n} = \frac{n}{2^{n-1}}}$, where n is any positive integer.

Hint: From Fig. 8-5, $A_1A_2 = 2R\sin\frac{\pi}{n}$, $A_1A_3 = 2R\sin\frac{2\pi}{n}, \cdots$, $A_1A_n = 2R\sin\frac{(n-1)\pi}{n}$. Multiplying term wise and making use of the result obtained in Example 8-9, the desired result follows.

8-16) Making use of the result proved in the previous problem, show that

$$\prod_{n=1}^{90} \sin n^{\circ} = \frac{3\sqrt{10}}{2^{89}}.$$

8-17) What is the sum of the following series

a) $1 + \frac{(\cos\phi)^2}{2!} + \frac{(\cos\phi)^4}{4!} + \frac{(\cos\phi)^6}{6!} + \frac{(\cos\phi)^8}{8!} + \frac{(\cos\phi)^{10}}{10!} + \cdots$

b) $\sinh\phi - \frac{(\sinh\phi)^3}{3!} + \frac{(\sinh\phi)^5}{5!} - \frac{(\sinh\phi)^7}{7!} + \frac{(\sinh\phi)^9}{9!} - \frac{(\sinh\phi)^{11}}{11!} + \cdots$

Hint: See (8-22), (8-23), (8-24) and (8-25).

(**Answer:** **a)** $\cosh(\cos\phi)$ **b)** $\sin(\sinh\phi)$).

8-18) If z and w are complex numbers, show that

a) $(\cosh z \pm \sinh z)^n = \cosh(nz) \pm \sinh(nz), \quad n = 0,1,2,3,\cdots$

b) $\tanh(z+w) = \frac{\tanh z + \tanh w}{1 + \tanh z \tanh w}$, and

c) $\tanh(2z) = \frac{2\tanh z}{1+(\tanh z)^2}$.

8-19) If $z = 1 - i\sqrt{3}$, show that

a) $z^6 = 64$, **b)** $z^8 + 128\bar{z} = 0$, **c)** $(\bar{z})^3 = -8$.

8-20) Find the real and the imaginary parts of the complex number $\sin(\cos\phi + i\sin\phi)$, where $\phi \in \mathbb{R}$.

Hint: See Problem 8-4.

9. Series of Complex Numbers.

a) The reader is supposed to be familiar with the Theory of sequences of complex numbers, convergence in the complex plane, etc, (see our e-book, **Sequences of Real and Complex Numbers**, by Demetrios P. Kanoussis, chapter 15).

b) Let $(z_n) = \{z_1,\ z_2, z_3, \cdots, z_k,\ z_{k+1},\ \cdots\}$ be a sequence **of complex numbers**, and let us form the **sequence of partial sums**,

$$s_1 = z_1,$$

$$s_2 = z_1 + z_2,$$

$$s_3 = z_1 + z_2 + z_3, \qquad (9\text{-}1)$$

$$\vdots$$

$$s_n = z_1 + z_2 + z_3 + \cdots + z_n,$$

$$\vdots$$

If the sequence (s_n) of partial sums converges to a limit z, we say that the series with complex terms $\sum_{n=1}^{\infty} z_n$ converges to z, and we write,

$$\lim_{n\to\infty} s_n = z_1 + z_2 + z_3 + \cdots + z_k + z_{k+1} + \cdots = \sum_{n=1}^{\infty} z_n = z. \qquad (9\text{-}2)$$

The following two Theorems are important in the investigation of series with complex terms.

Theorem 9-1.

If $z_n = x_n + iy_n$ where $x_n \in \mathbb{R}$ and $y_n \in \mathbb{R}$, then

$$\sum_{n=1}^{\infty} z_n = \sum_{n=1}^{\infty}(x_n + iy_n) = z = x + iy \Leftrightarrow$$

$$\sum_{n=1}^{\infty} x_n = x \ \text{ and } \ \sum_{n=1}^{\infty} y_n = y. \qquad (9\text{-}3)$$

Proof: For the proof see Example 9-1.

Theorem 9-2.

If $\sum z_n$ converges, then $\lim z_n = 0$.

Proof: For the proof see Example 9-2.

Note: This Theorem actually provides **a necessary condition for the convergence of $\sum z_n$.**

If $\lim z_n \neq 0\ (0 + i0)$, then $\sum z_n$ does not converge to a finite complex number z.

c) The series $\sum z_n = z_1 + z_2 + z_3 + \cdots$ is called **absolutely convergent, if the series $\sum |z_n| = |z_1| + |z_2| + |z_3| + \cdots$ is convergent.**

We note that, while **the series $\sum z_n$ is a series with complex terms, the series $\sum |z_n|$ is a series with real positive terms, and therefore all Theorems and Techniques developed in Chapter 5 (Convergence criteria for series with positive terms), can be applied for the investigation of the series $\sum |z_n|$.**

The following Theorem is of extreme importance.

Theorem 9-3.

If the series with real positive terms $\sum |z_n|$ converges, then the series with complex terms $\sum z_n$ converges as well. (Compare with Theorem 7-1).

Proof: Let $z_n = x_n + iy_n$ where $x_n \in \mathbb{R}$ and $y_n \in \mathbb{R}$.

The modulus of z_n is $|z_n| = \sqrt{{x_n}^2 + {y_n}^2} > 0$. By assumption, $\sum |z_n|$ converges, i.e. $\sum \sqrt{{x_n}^2 + {y_n}^2} < \infty$. Since $0 < |x_n| < \sqrt{{x_n}^2 + {y_n}^2}$, application of Theorem 5-1, (**the Comparison Test**), implies that $\sum |x_n| < \infty$, and therefore $\sum x_n < \infty$, **since absolute convergence implies convergence**,(Theorem 7-1), which in turn means that $\sum \boldsymbol{x_n} = \boldsymbol{x}$, $\boldsymbol{x}$ **being a finite real number**.

Reasoning similarly, we can prove that $\sum \boldsymbol{y_n} = \boldsymbol{y}$, where $\boldsymbol{y}$ **is also a finite real number**. Then the series

$$\sum z_n = \sum (x_n + iy_n) = \sum x_n + i \sum y_n = x + iy = z,$$

and this completes the proof.

Note: Theorem 9-3 reduces **the investigation of series with complex terms to the investigation of series with positive terms**, and therefore we may apply all Tests and Theorems developed in Chapter 5.

d) The exponential function e^z, $z \in \mathbb{C}$.

As a characteristic application of Theorem 9-3, we shall show that **the infinite series in (8-26) converges for any complex number z, and the sum of this series is by definition the function e^z.**

Let $\boldsymbol{z = r(\cos\phi + i\sin\phi)}$ be a complex number **in polar form**, (see (8-8)). If n is any positive integer, then by virtue of the **De Moivre's Theorem**

$$z^n = r^n(\cos(n\phi) + i\sin(n\phi)), \text{ and } |z^n| = |z|^n = r^n, \qquad (*)$$

since $|\cos(n\phi) + i\sin(n\phi)| = \sqrt{(\cos(n\phi))^2 + (\sin(n\phi))^2} = 1.$

Let us further consider the series with **complex terms**,

$$\sum_{n=0}^{\infty} \frac{z^n}{n!} = 1 + \frac{z}{1!} + \frac{z^2}{2!} + \frac{z^3}{3!} + \frac{z^4}{4!} + \cdots. \qquad (9\text{-}4)$$

The series in (9-4) **is absolutely convergent**, since

$$\sum_{n=0}^{\infty} \left|\frac{z^n}{n!}\right| = \sum_{n=0}^{\infty} \frac{|z^n|}{n!} = \sum_{n=0}^{\infty} \frac{|z|^n}{n!} = \sum_{n=0}^{\infty} \frac{|z|^n}{n!} = \sum_{n=0}^{\infty} \frac{r^n}{n!}, \text{ (from (*))},$$

and the series $\sum_{n=0}^{\infty} \frac{r^n}{n!} = e^r$, $\forall r \in \mathbb{R}$, (see Example 5-2 and Problem 7-17).

By Theorem 9-3, **the series in (9-4) is convergent for any** $z \in \mathbb{C}$, and the sum of this series is by definition the exponential function e^z, i.e.

$$\boldsymbol{e^z} = \sum_{n=0}^{\infty} \frac{z^n}{n!} = \mathbf{1} + \frac{z}{1!} + \frac{z^2}{2!} + \frac{z^3}{3!} + \frac{z^4}{4!} + \frac{z^5}{5!} + \cdots, \qquad z \in \mathbb{C}. \qquad (9\text{-}5)$$

We note that $\boldsymbol{e^z}$ **is a periodic function with fundamental period $2\pi i$**, since

$$e^{z+2\pi i} = e^z e^{2\pi i} = e^z(\cos(2\pi) + i\sin(2\pi)) = e^z(1 + i0) = e^z \quad (9\text{-}6)$$

In contrast to this, **the real exponential function e^x, $x \in \mathbb{R}$, is not periodic.**

e) Summation of real series of the form $\sum c_n \cos(n\phi)$ or $\sum c_n \sin(n\phi)$, where the coefficients c_n and the argument ϕ are all real numbers.

Several times we are confronted with the problem of finding **the sum of the series $\sum c_n \cos(n\phi)$ or $\sum c_n \sin(n\phi)$, in closed form.**

The following Theorem may be of great help towards this direction.

Theorem 9-4.

If the series $\sum c_n$ is absolutely convergent, then the series $\sum c_n \cos(n\phi)$ and $\sum c_n \sin(n\phi)$, are convergent, for any real ϕ.

Proof: By assumption $\sum|c_n| < \infty$ and since $|c_n \cos(n\phi)| \leq |c_n|$ and $|c_n \sin(n\phi)| \leq |c_n|$, Theorem 5-1 implies that $\sum|c_n \cos(n\phi)|$ and $\sum|c_n \sin(n\phi)|$ are convergent, and therefore, by Theorem 7-1, $\sum c_n \cos(n\phi)$ **and $\sum c_n \sin(n\phi)$ are convergent** as well, and this completes the proof.

Suppose now that **we want to evaluate the sum of the series $\sum c_n \cos(n\phi)$**, where $\sum|c_n| < \infty$. In order to do that, we may follow **the general procedure**, outlined below.

Let

$$X = \sum_{n=1}^{\infty} c_n \cos(n\phi) = c_1 \cos\phi + c_2 \cos 2\phi + c_3 \cos 3\phi + \cdots \quad (*)$$

At the same time we consider the corresponding series,

$$Y = \sum_{n=1}^{\infty} c_n \sin(n\phi) = c_1 \sin\phi + c_2 \sin 2\phi + c_3 \sin 3\phi + \cdots \quad (**)$$

and we form the sum,

$$X + iY = c_1(\cos\phi + i\sin\phi) + c_2(\cos 2\phi + i\sin 2\phi) +$$
$$+c_3(\cos 3\phi + i\sin 3\phi) + \cdots,$$

or making use of **the Euler's formula,**

$$X + iY = c_1 e^{i\phi} + c_2 e^{i2\phi} + c_3 e^{i3\phi} + \cdots = \sum_{n=1}^{\infty} c_n e^{in\phi}.$$

If the sum of the series $\sum_{n=1}^{\infty} c_n e^{in\phi}$ can be found, in closed form, say $S_1 + iS_2$, then

$$X = \sum_{n=1}^{\infty} c_n \cos(n\phi) = S_1 \text{ and } Y = \sum_{n=1}^{\infty} c_n \sin(n\phi) = S_2. \quad (9\text{-}7)$$

Thus, using complex numbers we not only find the series of the cosine terms, (eq. (*)), **but as an extra bonus**, we find the sum of the corresponding series, with the sine terms.

The following examples, illustrate the general procedure, outlined above.

Example 9-1.

Prove Theorem 9-1.

Solution

i) Let us assume that $\sum_{n=1}^{\infty} z_n = \sum_{n=1}^{\infty}(x_n + iy_n) = x + iy = z$. Then,

$(x_1 + iy_1) + (x_2 + iy_2) + (x_3 + iy_3) + \cdots = x + iy$, or

$(x_1 + x_2 + x_3 + \cdots) + i(y_1 + y_2 + y_3 + \cdots)x + iy$, i.e.
$\sum_{n=1}^{\infty} x_n + i\sum_{n=1}^{\infty} y_n = x + iy$, therefore $\sum_{n=1}^{\infty} \boldsymbol{x_n} = \boldsymbol{x}$ and $\sum_{n=1}^{\infty} \boldsymbol{y_n} = \boldsymbol{y}$ according to (8-1), (equality between two complex numbers).

ii) If $\sum_{n=1}^{\infty} x_n = x$ and $\sum_{n=1}^{\infty} y_n = y$, then

$\sum_{n=1}^{\infty} x_n + i\sum_{n=1}^{\infty} y_n = x + iy$, or $\sum_{n=1}^{\infty}(x_n + iy_n) = x + iy$, or

$\sum_{n=1}^{\infty} z_n = z$, where $z_n = x_n + iy_n$ and $z = x + iy$.

Example 9-2.

Prove Theorem 9-2.

Solution

If (s_n) is the sequence of partial sums of **a convergent series** $\sum \boldsymbol{z_n} = \boldsymbol{z}$, then, as we know, $\boldsymbol{\lim_{n\to\infty} s_n = z}$. The general term z_n of the series, is $\boldsymbol{z_n = s_n - s_{n-1}}$, and passing to the limit as $n \to \infty$, we have,

$\lim_{n\to\infty} z_n = \lim_{n\to\infty}(s_n - s_{n-1}) = \lim_{n\to\infty} s_n - \lim_{n\to\infty} s_{n-1} = z - z = 0.$

Example 9-3.

a) If z is any complex number ($\boldsymbol{z \neq 1 = 1 + i0}$), show that

$$\boldsymbol{1 + z + z^2 + z^3 + z^4 + \cdots + z^n = \frac{z^{n+1} - 1}{z - 1}}.$$

b) If $|z| < 1$, show that $\sum_{n=0}^{\infty} z^n = 1 + z + z^2 + z^3 + z^4 + \cdots = \frac{1}{1-z}$.

Solution

a) Let $S_n = 1 + z + z^2 + z^3 + \cdots + z^n$. Then,

$zS_n = z + z^2 + z^3 + \cdots + z^n + z^{n+1}$, and subtracting from the first equation, we obtain, $(1-z)S_n = 1 - z^{n+1}$, or $\boldsymbol{S_n = \frac{1-z^{n+1}}{1-z}}$.

b) If $|z| < 1$ then $\lim z^n = 0$, and

$$S = \lim S_n = \lim \left(\frac{1-z^{n+1}}{1-z}\right) = \frac{1-0}{1-z} = \frac{1}{1-z}.$$

Example 9-4.

If $0 < r < 1$, find $\sum_{n=1}^{\infty} r^n \cos(n\theta)$ and $\sum_{n=1}^{\infty} r^n \sin(n\theta)$, $\theta \in \mathbb{R}$.

Solution

Let $\boldsymbol{X = \sum_{n=1}^{\infty} r^n \cos(n\theta)}$ and $\boldsymbol{Y = \sum_{n=1}^{\infty} r^n \sin(n\theta)}$. Then,

$$X + iY = \sum_{n=1}^{\infty} r^r(\cos(n\theta) + i\sin(n\theta)) = \sum_{n=1}^{\infty} r^n e^{in\theta} = \sum_{n=1}^{\infty} (re^{i\theta})^n. \quad (*)$$

Since $|re^{i\theta}| = r < 1$, from Example 9-3(b), we have,

$$\sum_{n=1}^{\infty} (re^{i\theta})^n = \frac{1}{1-re^{i\theta}} - 1 = \frac{re^{i\theta}}{1-re^{i\theta}} = \frac{r(\cos\theta + i\sin\theta)}{1-r(\cos\theta+i\sin\theta)} \quad \text{or,}$$

$$\sum_{n=1}^{\infty} (re^{i\theta})^n = \frac{r\cos\theta + ir\sin\theta}{(1-r\cos\theta) - ir\sin\theta} = \frac{(r\cos\theta+ir\sin\theta)\{(1-r\cos\theta)+ir\sin\theta\}}{(1-r\cos\theta)^2+(r\sin\theta)^2},$$

which is easily simplified to the following,

$$X + iY = \sum_{n=1}^{\infty} (re^{i\theta})^n = \frac{r\cos\theta - r^2}{1+r^2-2r\cos\theta} + i\frac{r\sin\theta}{1+r^2-2r\cos\theta},$$

from which one easily finds,

$$\boldsymbol{X = \sum_{n=1}^{\infty} r^n \cos(n\theta) = \frac{r\cos\theta - r^2}{1+r^2-2r\cos\theta}} \quad \text{and}$$

$$\boldsymbol{Y = \sum_{n=1}^{\infty} r^n \sin(n\theta) = \frac{r\sin\theta}{1+r^2-2r\cos\theta}}.$$

Example 9-5.

Find the sum of the series $\sum_{n=0}^{\infty} \frac{\cos(n\theta)}{n!}$.

Solution

The given series is obviously convergent, (why?). In order to find the sum the series, we consider in parallel to the given series, the corresponding series with sine terms, $\sum_{n=0}^{\infty} \frac{\sin(n\theta)}{n!}$ which is also convergent.

Let $\boldsymbol{X} = \sum_{\boldsymbol{n}=\boldsymbol{0}}^{\infty} \frac{\cos(\boldsymbol{n\theta})}{\boldsymbol{n}!}$ and $\boldsymbol{Y} = \sum_{\boldsymbol{n}=\boldsymbol{0}}^{\infty} \frac{\sin(\boldsymbol{n\theta})}{\boldsymbol{n}!}$. Then,

$$X + iY = \sum_{n=0}^{\infty} \frac{\cos(n\theta)+i\sin(n\theta)}{n!} = \sum_{n=0}^{\infty} \frac{e^{in\theta}}{n!} = \sum_{n=0}^{\infty} \frac{\left(e^{i\theta}\right)^n}{n!},$$

and since

$$\sum_{n=0}^{\infty} \frac{\left(e^{i\theta}\right)^n}{n!} = e^{e^{i\theta}} = e^{(\cos\theta + i\sin\theta)} = e^{\cos\theta} e^{i\sin\theta} =$$
$$= e^{\cos\theta}(\cos(\sin\theta) + i\sin(\sin\theta)),$$

equating real and imaginary parts yields

$$\boldsymbol{X} = \sum_{\boldsymbol{n}=\boldsymbol{0}}^{\infty} \frac{\cos(\boldsymbol{n\theta})}{\boldsymbol{n}!} = \boldsymbol{e}^{\cos\boldsymbol{\theta}} \cos(\sin\boldsymbol{\theta}), \text{ and}$$

$$\boldsymbol{Y} = \sum_{\boldsymbol{n}=\boldsymbol{0}}^{\infty} \frac{\sin(\boldsymbol{n\theta})}{\boldsymbol{n}!} = \boldsymbol{e}^{\cos\boldsymbol{\theta}} \sin(\sin\boldsymbol{\theta}).$$

PROBLEMS

9-1) Find the sum of the series $\frac{\sin 2\theta}{2!} + \frac{\sin 4\theta}{4!} + \frac{\sin 6\theta}{6!} + \frac{\sin 8\theta}{8!} + \cdots \quad \theta \in \mathbb{R}$.

Hint: Making use of (8-31) and (9-4), show that

$$\cosh \boldsymbol{z} = \boldsymbol{1} + \frac{\boldsymbol{z}^2}{2!} + \frac{\boldsymbol{z}^4}{4!} + \frac{\boldsymbol{z}^6}{6!} + \frac{\boldsymbol{z}^8}{8!} + \cdots, \text{ and } \sinh \boldsymbol{z} = \boldsymbol{z} + \frac{\boldsymbol{z}^3}{3!} + \frac{\boldsymbol{z}^5}{5!} + \frac{\boldsymbol{z}^7}{7!} + \frac{\boldsymbol{z}^9}{9!} + \cdots$$

If $\boldsymbol{z} = \boldsymbol{x} + \boldsymbol{i0} = \boldsymbol{x} \in \mathbb{R}$, these formulas reduce to (8-22) and (8-23), respectively. Also make use of (8-34) and (8-35).

(**Answer:** $\sinh(\cos\theta)\sin(\sin\theta)$).

9-2) Find the sum of the series

$$1+\frac{\cos 2\theta}{2!}+\frac{\cos 4\theta}{4!}+\frac{\cos 6\theta}{6!}+\frac{\cos 8\theta}{8!}+\cdots \quad \theta \in \mathbb{R}.$$

9-3) Find the sum of the series $\frac{\sin \theta}{1!}+\frac{\sin 2\theta}{2!}+\frac{\sin 3\theta}{3!}+\frac{\sin 4\theta}{4!}+\cdots \quad \theta \in \mathbb{R}.$

(Answer: $e^{\cos\theta}\sin(\sin\theta)$).

9-4) Show that for any real θ,

$$\sin\theta-\frac{\sin 3\theta}{3!}+\frac{\sin 5\theta}{5!}-\frac{\sin 7\theta}{7!}+\cdots=\cos(\cos\theta)\sinh(\sin\theta).$$

9-5) Find the **finite sums,**

a) $x_n=\sum_{k=0}^{n} r^k\cos(\theta+k\phi)$, and **b)** $y_n=\sum_{k=0}^{n} r^k\sin(\theta+k\phi)$,
($r\neq 0,\ \theta\in\mathbb{R},\ \phi\in\mathbb{R}$).

(Answer: $x_n=\frac{\cos\theta-r\cos(\theta-\phi)-r^{n+1}\cos\{(n+1)\phi+\theta\}+r^{n+2}\cos(n\phi+\theta)}{1+r^2-2r\cos\phi}$,

$y_n=\frac{\sin\theta-r\sin(\theta-\phi)-r^{n+1}\sin\{(n+1)\phi+\theta\}+r^{n+2}\sin(n\phi+\theta)}{1+r^2-2r\cos\phi}$.)

9-6) If $|r|<1$, ($r\neq 0$), show that

$$\sum_{k=0}^{\infty} r^k\cos(\theta+k\phi)=\frac{\cos\theta-r\cos(\theta-\phi)}{1+r^2-2r\cos\phi},\quad \text{and}$$

$$\sum_{k=0}^{\infty} r^k\sin(\theta+k\phi)=\frac{\sin\theta-r\sin(\theta-\phi)}{1+r^2-2r\cos\phi}.$$

9-7) If n is any positive integer, show that,

$$\sum_{k=1}^{n-1}\sin\left(k\frac{\pi}{n}\right)=\cot\left(\frac{\pi}{2n}\right),\quad \text{and}\quad \sum_{k=1}^{n-1}\cos\left(k\frac{2\pi}{n}\right)=0.$$

9-8) If ϕ is a real number, find in closed form the **finite sums,**

$$S_1=\sum_{k=1}^{n}\sin\{(2k-1)\phi\},\quad \text{and}\quad S_2=\sum_{k=1}^{n}\cos\{(2k-1)\phi\}.$$

9-9) Using the results of Problem 9-8, show that $\tan(n\phi)=\frac{\sum_{k=1}^{n}\sin\{(2k-1)\phi\}}{\sum_{k=1}^{n}\cos\{(2k-1)\phi\}}$.

9-10) Investigate the convergence of the following series,

a) $\sum \frac{1}{(1+i)^n}$ and **b)** $\sum \frac{\cos\left(\frac{n\pi}{5}\right)+i\sin\left(\frac{n\pi}{5}\right)}{(1+i)^n}$.

Hint: Both series are **absolutely convergent**, (why?), therefore by Theorem 9-3, they are both convergent. To show for example that $\sum \frac{1}{|1+i|^n} = \sum \frac{1}{(\sqrt{2})^n}$ is convergent, we may apply the **D' Alembert's Test**, etc.

9-11) Consider the series $\sum_{k=1}^{\infty} \frac{k^k}{k!} z^k$, $z \in \mathbb{C}$, and show that it converges for $|z| < \frac{1}{e}$.

9-12) If $|z| < \sqrt{3}$, $(z \in \mathbb{C})$, show that $\sum_{k=0}^{\infty} \frac{(-1)^k}{3^{2(k+1)}} z^{4k+1} = \frac{z}{z^4+9}$.

Hint: See Example 9-3.

9-13) Making use of (9-5), (8-31) and (8-32), show that if z is any complex number, then

$$\cosh z = \sum_{n=0}^{\infty} \frac{z^{2n}}{(2n)!}, \quad \text{and} \quad \sinh z = \sum_{n=0}^{\infty} \frac{z^{2n+1}}{(2n+1)!}.$$

9-14) Find the sum of the series $\sum_{n=0}^{\infty} \frac{(z+2\pi i)^{2n}}{(2n)!}$.

Hint: Make use of Problem 9-13.

9-15) a) For which values of $z \in \mathbb{C}$, the following series is absolutely convergent?

$$(z-i) + \frac{(z-i)^2}{2} + \frac{(z-i)^3}{3} + \frac{(z-i)^4}{4} + \cdots$$

b) Sketch the region of absolute convergence.

(**Answer: a)** $|z-i| < 1$ **b)** The interior of a circle, having center at $(0, i)$ and radius 1).

10. Multiple Series.

A finite sum with two indices of summation, for example

$$\sum_{i=1}^{3}\sum_{j=1}^{2} u_{ij} = \sum_{i=1}^{3}(u_{i1} + u_{i2}) = u_{11} + u_{12} + u_{21} + u_{22} + u_{31} + u_{32},$$

is called a double finite sum.

In general, we may have a double finite sum, of the form,

$$\sum_{i=1}^{m}\sum_{j=1}^{n} u_{ij}, \qquad (10\text{-}1)$$

To compute this sum, we must perform summation with respect to j, ($1 \leq j \leq n$), keeping i fixed, and then perform the summation with respect to i, ($1 \leq i \leq m$). In other words,

$$\sum_{i=1}^{m}\sum_{j=1}^{n} u_{ij} = \sum_{i=1}^{m}(u_{i1} + u_{i2} + \cdots + u_{in}) = (u_{11} + u_{12} + \cdots + u_{1n}) + (u_{21} + u_{22} + \cdots + u_{2n}) + \cdots + (u_{m1} + u_{m2} + \cdots + u_{mn}). \qquad (10\text{-}2)$$

It is obvious that **the double finite sum (10-2) represents the sum of all the elements of the $m \times n$ matrix,**

$$\begin{matrix} u_{11} & u_{12} & u_{13} & \cdots & u_{1n} \\ u_{21} & u_{22} & u_{23} & \cdots & u_{2n} \\ \vdots & \vdots & \vdots & \vdots & \vdots \\ u_{m1} & u_{m2} & u_{m3} & \cdots & u_{mn} \end{matrix}$$

Likewise, we may have **a double infinite series**, of the form,

$$\sum_{i=1}^{\infty}\sum_{j=1}^{\infty} u_{ij}. \qquad (10\text{-}3)$$

We shall consider the simplest case, where **all $u_{ij} \geq 0$**.

The double series (10-3) thus represents the sum,

$$u_{11} + u_{12} + u_{21} + u_{13} + u_{22} + u_{31} + u_{14} + u_{23} + u_{32} + u_{41} + \cdots \qquad (10\text{-}4)$$

which can be considered as **a series of single terms**.

The sum (10-4) **either converges to a finite positive number, or diverges to** $+\infty$**,** i.e. if all $u_{ij} \geq 0$, then

either $\sum_{i=1}^{\infty}\sum_{j=1}^{\infty} \boldsymbol{u_{ij}} < \infty$, or $\sum_{i=1}^{\infty}\sum_{j=1}^{\infty} \boldsymbol{u_{ij}} = +\infty$. (10-5)

In the first case (**convergence**), the sum of the series does not depend on the order of the summands. In other words, **any rearrangement of the positive terms in (10-4), does not affect the sum S of the series**, (see Theorem 7-4). We may thus write,

$$\sum_{i=1}^{\infty}\sum_{j=1}^{\infty} \boldsymbol{u_{ij}} = \sum_{i=1}^{\infty}\left(\sum_{j=1}^{\infty} \boldsymbol{u_{ij}}\right) = \sum_{j=1}^{\infty}\left(\sum_{i=1}^{\infty} \boldsymbol{u_{ij}}\right) = \boldsymbol{S}. \qquad (10\text{-}6)$$

For arbitrary terms u_{ij} (**terms of any sign or complex terms**), the series (10-3), is called **absolutely convergent**, if

$$\sum_{i=1}^{\infty}\sum_{j=1}^{\infty}\left|\boldsymbol{u_{ij}}\right| < \infty. \qquad (10\text{-}7)$$

The following two Theorems, which we state without proof, are of particular importance.

Theorem 10-1.

Any absolutely convergent double series is itself convergent, i.e.

If $\sum_{i=1}^{\infty}\sum_{j=1}^{\infty}\left|\boldsymbol{u_{ij}}\right| < \infty$**, then** $\sum_{i=1}^{\infty}\sum_{j=1}^{\infty} \boldsymbol{u_{ij}} < \infty$**.**

Theorem 10-2.

If $\boldsymbol{S_1} = \sum_{i=1}^{\infty} \boldsymbol{x_i}$ **and** $\boldsymbol{S_2} = \sum_{i=1}^{\infty} \boldsymbol{y_i} = \sum_{j=1}^{\infty} \boldsymbol{y_j}$ **be any two absolutely convergent series, then**

$$\boldsymbol{S_1 S_2} = \left(\sum_{i=1}^{\infty} \boldsymbol{x_i}\right)\left(\sum_{j=1}^{\infty} \boldsymbol{y_j}\right) = \sum_{i=1}^{\infty}\left(\boldsymbol{x_i}\sum_{j=1}^{\infty} \boldsymbol{y_j}\right) = \sum_{i=1}^{\infty}\left(\sum_{j=1}^{\infty} \boldsymbol{x_i y_j}\right) = \sum_{i=1}^{\infty}\sum_{j=1}^{\infty} \boldsymbol{x_i y_j}. \qquad (10\text{-}8)$$

Thus, **multiplication of two absolutely convergent series, results in an absolutely convergent double series.**

Example 10-1.

Let us consider **the infinite array** of real numbers,

$$\begin{matrix} u_{11} & u_{12} & u_{13} & \cdots & u_{1n} & \cdots \\ u_{21} & u_{22} & u_{23} & \cdots & u_{2n} & \cdots \\ \vdots & \vdots & \vdots & \vdots & \vdots & \vdots \\ u_{m1} & u_{m2} & u_{m3} & \cdots & u_{mn} & \cdots \\ \vdots & \vdots & \vdots & \vdots & \vdots & \vdots \end{matrix}$$

If all $\boldsymbol{u_{ij}}$ **are positive numbers,** and $\sum_{i=1}^{\infty}\sum_{j=1}^{\infty} u_{ij} < \infty$, show that $\sum_{j=1}^{\infty} u_{mj} < \infty$, $(m = 1,2,3,\cdots)$, and $\sum_{k=1}^{\infty} u_{kr} < \infty$, $(r = 1,2,3,\cdots)$.

In other words, **if a double series of positive numbers is convergent, then the series of all positive terms belonging to any particular row (or column), will be convergent as well.**

Solution

By assumption, $\sum_{i=1}^{\infty}\sum_{j=1}^{\infty} u_{ij} < \infty$, $(\boldsymbol{u_{ij}} > 0)$, meaning that $\sum_{i=1}^{\infty}\sum_{j=1}^{\infty} u_{ij} = S$, a finite positive number, i.e.

$$(u_{11} + u_{12} + u_{13} +\cdot\cdot) + (u_{21} + u_{22} + u_{23} +\cdot\cdot) + \cdots +$$

$$+(u_{m1} + u_{m2} + u_{m3} +\cdot\cdot) + \cdots = S \qquad (*)$$

The series formed by the elements of the $m^{\underline{th}}$ row is

$$(u_{m1} + u_{m2} + u_{m3} + \cdots).$$

The sequence of the partial sums of this series,

$$\{u_{m1}, \quad u_{m1} + u_{m2}, \quad u_{m1} + u_{m2} + u_{m3}, \quad u_{m1} + u_{m2} + u_{m3} + u_{m4}, \cdots\}$$

is increasing and bounded above by the number $\boldsymbol{S}$, therefore this series converges to **a finite number** $\boldsymbol{S_m \leq S}$, $(\boldsymbol{m = 1, 2, 3, \cdots})$.

In a similar manner, we may show the convergence of the series formed by the elements of the $n^{\underline{th}}$ column.

Example 10-2.

Find the sum of the series $\sum_{m=1}^{\infty}\sum_{n=1}^{\infty}\frac{1}{m^2n^2}$.

Solution

Application of Theorem 10-2, yields

$$\sum_{m=1}^{\infty}\sum_{n=1}^{\infty}\frac{1}{m^2n^2}=\left(\sum_{m=1}^{\infty}\frac{1}{m^2}\right)\left(\sum_{n=1}^{\infty}\frac{1}{n^2}\right)=\frac{\pi^2}{6}\cdot\frac{\pi^2}{6}=\frac{\pi^4}{36}.$$

(See Note in Example 5-3)

PROBLEMS

10-1) Show that $\sum_{m=1}^{\infty}\sum_{n=1}^{\infty}\frac{1}{m^4n^4}=\frac{\pi^8}{8100}$ and $\sum_{m=1}^{\infty}\sum_{n=1}^{\infty}\frac{1}{m^2n^4}=\frac{\pi^6}{540}$.

Hint: See Note in Example 5-3.

10-2) If n is any integer, (positive, negative or zero), one may define the so called **Bessel function of the first kind and order n, $J_n(x)$,** by means of the infinite series,

$$J_n(x)=\sum_{k=0}^{\infty}\frac{(-1)^k}{k!(n+k)!}\left(\frac{x}{2}\right)^{n+2k}.$$

a) Making use of Theorem 7-3, show that the series is absolutely convergent, and therefore convergent for all values of $x\in\mathbb{R}$.

b) Show that

$$J_0(x)=1-\frac{x^2}{2^2}+\frac{x^4}{2^24^2}-\frac{x^6}{2^24^26^2}+\cdots.$$

Note: Bessel's functions appear in a natural way, in Applied Mathematics, while solving certain types of **Partial Differential Equations**,(for instance the Wave or the Heat-Diffusion Equation), **in Cylindrical coordinates**.

10-3) If $t \neq 0$, show that,

a) $e^{\frac{x}{2}\left(t-\frac{1}{t}\right)} = \sum_{k=0}^{\infty} \sum_{m=0}^{\infty} \frac{(-1)^m \left(\frac{x}{2}\right)^{k+m}}{k!m!} t^{k-m}$ (*)

b) If we set $n = k - m$, in which case $\boldsymbol{n}$ **varies between** $-\infty$ **and** $+\infty$, (i.e. $n \in \{\cdots, -2, -1, 0, 1, 2, \cdots\}$), show that (*) can be expressed as

$e^{\frac{x}{2}\left(t-\frac{1}{t}\right)} = \sum_{n=-\infty}^{\infty} J_n(x)\, t^n.$ (**)

Note: Due to (**), the function $e^{\frac{x}{2}\left(t-\frac{1}{t}\right)}$ is known as **the Bessel's generating function**.

10-4) Setting $t = e^{i\theta}$, where $i = \sqrt{-1}$, $\theta \in \mathbb{R}$, and making use of **the Euler's formula** and **the De Moivre's Theorem**, show that,

a) $e^{ix \sin\theta} = \sum_{n=-\infty}^{\infty} J_n(x) e^{in\theta}$, and

b) $\cos(x \sin\theta) = J_0(x) + 2\{J_2(x)\cos 2\theta + J_4(x)\cos 4\theta + J_6(x)\cos 6\theta + \cdots\}$,

$\sin(x \sin\theta) = 2\{J_1(x)\sin\theta + J_3(x)\sin 3\theta + J_5(x)\sin 5\theta + \cdots\}$.

Hint: Using the infinite series expression for $J_n(x)$, in Problem 10-2, show first that $\boldsymbol{J_{-n}(x) = (-1)^n J_n(x)}$, where $\boldsymbol{n}$ **is any nonnegative integer number**.

11. Infinite Products.

Let $(u_n) = \{u_1, u_2, u_3, \cdots, u_n, u_{n+1}, \cdots\}$ be a given sequence of Real or Complex numbers, where we assume that $\boldsymbol{u_k \neq 0}, \ \forall \boldsymbol{k}$. From the given sequence, we may form a new sequence $(P_n) = \{P_1, P_2, P_3, \cdots, P_n, \cdots\}$, in the following way,

$$\boldsymbol{P_1 = u_1}$$
$$\boldsymbol{P_2 = u_1 u_2 = \prod_{k=1}^{2} u_k}$$
$$\boldsymbol{P_3 = u_1 u_2 u_3 = \prod_{k=1}^{3} u_k}$$
$$\cdots \quad \cdots \quad \cdots \quad \cdots$$
$$\boldsymbol{P_n = u_1 u_2 u_3 \cdots u_n = \prod_{k=1}^{n} u_k}$$
$$\cdots \quad \cdots \quad \cdots \quad \cdots \quad \cdots \quad \cdots$$

The sequence (P_n) is called the sequence of the partial products of (u_n). The number P_n is called the $n^{\underline{th}}$ partial product of the sequence (P_n).

Obviously,

$$\lim_{n\to\infty} \boldsymbol{P_n} = \lim_{n\to\infty}\left(\prod_{k=1}^{n} u_k\right) = \prod_{k=1}^{\infty} u_k = u_1 u_2 u_3 \cdots u_n u_{n+1} \cdots \qquad (11\text{-}1)$$

The expression $\prod_{k=1}^{\infty} u_k$ is called **an infinite product**. **The term u_k is the $k^{\underline{th}}$ term or the general term of the infinite product**.

The **index k is a dummy index**, in the sense that

$\prod_{k=1}^{\infty} u_k = \prod_{n=1}^{\infty} u_n = \prod_{m=1}^{\infty} u_m$, etc,

since each one of these infinite products represents the product $u_1 u_2 u_3 \cdots u_k u_{k+1} \cdots$.

Definition 11-1: If the sequence (P_n) of the partial products, converges to **a finite number $P \neq 0$**, we say that **the infinite product converges to P**, and we write,

$$\lim \boldsymbol{P_n = P} \Leftrightarrow \prod_{n=1}^{\infty} u_n = \boldsymbol{P}. \qquad (11\text{-}2)$$

Otherwise we say that the infinite product **is divergent.**

In particular, **if $P = 0$, (while $u_k \neq 0 \quad \forall k$**), we say that **the infinite product diverges to 0**.

In general, the terms $u_1, u_2, u_3, \cdots$ could be **either numbers (real or complex), or functions of a real variable x or a complex variable z.**

For instance, if $u_n = u_n(x)$, $n = 1,2,3,\cdots$, then the infinite product $\prod_{n=1}^{\infty} u_n(x) = P(x)$, and like in series, **this product could converge for some values of x and diverge for the rest**.

As an example, it can be shown that the infinite product $\prod_{n=1}^{\infty}\left(1-\frac{x^2}{\pi^2 n^2}\right)$, converges for all values of $x \in \mathbb{R}$, and it represents the function $\frac{\sin x}{x}$, i.e.

$$\prod_{n=1}^{\infty}\left(1-\frac{x^2}{\pi^2 n^2}\right) = \frac{\sin x}{x}, \quad -\infty < x < \infty. \tag{11-3}$$

Note: If the general term u_n of a product can be expressed as $u_n = \frac{\phi_{n+1}}{\phi_n}$, (or as $u_n = \frac{\phi_n}{\phi_{n+1}}$), **where ϕ_n is a suitable function of n**, then the evaluation of the infinite product is greatly simplified, since

$$P_n = u_1 u_2 u_3 \cdots u_{n-1} u_n = \frac{\phi_2}{\phi_1} \cdot \frac{\phi_3}{\phi_2} \cdot \frac{\phi_4}{\phi_3} \cdots \frac{\phi_n}{\phi_{n-1}} \cdot \frac{\phi_{n+1}}{\phi_n} = \frac{\phi_{n+1}}{\phi_1}, \text{ and}$$

$$P = \lim P_n = \prod_{n=1}^{\infty} u_n = \lim\left(\frac{\phi_{n+1}}{\phi_1}\right) = \frac{1}{\phi_1} \cdot \lim \phi_{n+1}.$$

As an Example, let us consider the infinite product $\prod_{n=2}^{\infty}\left(1-\frac{1}{n^2}\right)$. We have,

$$P_k = \prod_{n=2}^{k}\left(1-\frac{1}{n^2}\right) = \prod_{n=2}^{k}\frac{n^2-1}{n^2} = \prod_{n=2}^{\infty}\frac{(n-1)(n+1)}{n^2}, \text{ or}$$

$$P_k = \frac{1\cdot 3}{2^2} \cdot \frac{2\cdot 4}{3^2} \cdot \frac{3\cdot 5}{4^2} \cdots \frac{(k-2)k}{(k-1)^2} \cdot \frac{(k-1)(k+1)}{k^2}, \text{ or}$$

$$P_k = \frac{1}{2} \cdot \frac{3}{2} \cdot \frac{2}{3} \cdot \frac{4}{3} \cdot \frac{3}{4} \cdot \frac{5}{4} \cdots \frac{k-2}{k-1} \cdot \frac{k}{k-1} \cdot \frac{k-1}{k} \cdot \frac{k+1}{k}, \text{or}$$

$$P_k = \frac{1}{2} \cdot \frac{k+1}{k} \Rightarrow \prod_{n=2}^{\infty}\left(1-\frac{1}{n^2}\right) = \lim_{k\to\infty} P_k = \lim_{k\to\infty}\left(\frac{1}{2} \cdot \frac{k+1}{k}\right) = \frac{1}{2} \cdot 1 = \frac{1}{2}.$$

The following Theorems are important, while investigating infinite products.

Theorem 11-1 (Necessary Condition for Convergence).

If the infinite product $\prod u_n$ converges to a finite number $P \neq 0$, then the general term u_n tends to 1, as $n \to \infty$. In symbols,

If $\prod u_n = P$, $(P \neq 0)$, then $\lim_{n\to\infty} u_n = 1$.

Proof: Let us assume that the infinite product converges to **a finite number $P \neq 0$**, i.e. $\prod u_n = P$. This means that $\lim u_n = P$. The general term u_n can be expressed as,

$$u_n = \frac{u_1 u_2 u_3 \cdots u_{n-1} u_n}{u_1 u_2 u_3 \cdots u_{n-1}} = \frac{P_n}{P_{n-1}}$$

and passing to the limit as $n \to \infty$, we have,

$$\lim_{n\to\infty} u_n = \lim_{n\to\infty} \frac{P_n}{P_{n-1}} = \frac{\lim_{n\to\infty} P_n}{\lim_{n\to\infty} P_{n-1}} = \frac{P}{P} = 1.$$

Note: If $\lim u_n \neq 1$, then the infinite product $\prod u_n$ does not converge. However, if $\lim u_n = 1$, the infinite product $\prod u_n$ may or may not converge.

As an Example, let us consider the $\prod_{n=1}^{\infty} \left(\frac{2n+3}{5n+1}\right)$. The general term $u_n = \frac{2n+3}{5n+1}$, and since $\lim u_n = \frac{2}{5} \neq 1$, **the given infinite product is not convergent.**

On the other hand, we consider the product $\prod_{n=1}^{\infty} \frac{n+1}{n}$, in which case $\lim \frac{n+1}{n} = 1$. However, **this product diverges to** $+\infty$, since,

$$P_k = \prod_{n=1}^{k} u_n = \frac{2}{1} \cdot \frac{3}{2} \cdot \frac{4}{3} \cdots \frac{k}{k-1} \cdot \frac{k+1}{k} = k+1, \text{ and therefore,}$$

$$\prod_{n=1}^{\infty} \frac{n+1}{n} = \lim_{k\to\infty} P_k = \lim_{k\to\infty}(k+1) = +\infty.$$

In conclusion, **if $\lim_{n\to\infty} u_n \neq 1$, then the infinite product $\prod_{n=1}^{\infty} u_n$ does not converge, while if $\lim_{n\to\infty} u_n = 1$, then the infinite product may either be convergent or may be divergent, (additional investigation is needed).**

(Compare with Theorem 4-5, Necessary condition for convergence of a series).

Let us now consider the $\prod_{n=1}^{\infty} u_n$ converging to a finite number $P \neq 0$. By virtue of Theorem 11-1, $\lim u_n = 1$, therefore we may set,

$$\boldsymbol{u_n = 1 + w_n} \text{ where } \boldsymbol{\lim w_n = 0}. \tag{11-4}$$

The infinite product can now be expresses as

$$\boldsymbol{\prod_{n=1}^{\infty}(1 + w_n) = P} \text{ where } \boldsymbol{\lim w_n = 0}. \tag{11-5}$$

In this case, the necessary condition for convergence, (Theorem 11-1), may be stated as follows:

If $\lim w_n \neq 0$, the infinite product $\prod_{n=1}^{\infty}(1 + w_n)$ does not converge.

For instance, the infinite products,

$\prod\left(1 + \frac{n+1}{n}\right)$, $\prod(1 + Arctan(n))$ and $\prod\left(1 + \sqrt[n]{n}\right)$ **are not convergent,** since,

$$\lim \frac{n+1}{n} = 1 \neq 0, \quad \lim(Arctan(n)) = \frac{\pi}{2} \neq 0, \quad \lim \sqrt[n]{n} = 1 \neq 0.$$

Definition 11-2: If the infinite product $\prod_{n=1}^{\infty}(1 + |w_n|)$ converges, we say that $\prod_{n=1}^{\infty}(1 + w_n)$ **is absolutely convergent.**

If $\prod_{n=1}^{\infty}(1 + w_n)$ converges, but $\prod_{n=1}^{\infty}(1 + |w_n|)$ diverges, we say that the infinite product $\prod_{n=1}^{\infty}(1 + w_n)$ **is conditionally convergent.**

The following important Theorems are frequently used, in order to investigate the nature of infinite products.

Theorem 11-2.

If $\sum_{n=1}^{\infty}|w_n|$ converges, then $\prod_{n=1}^{\infty}(1 + |w_n|)$ converges, and conversely, if $\prod_{n=1}^{\infty}(1 + |w_n|)$ converges, then $\sum_{n=1}^{\infty}|w_n|$ converges as well.

Proof: a) Let us assume that $\sum_{n=1}^{\infty}|w_n|$ **is convergent to a finite positive number w**, i.e. $\sum_{n=1}^{\infty}|w_n| = w > 0$. We shall show that $\prod_{n=1}^{\infty}(1 + |w_n|)$ **is convergent as well.**

Let us now consider the sequence (P_n) of the partial products, where,

$$P_n = (1 + |w_1|)(1 + |w_2|)(1 + |w_3|) \cdots (1 + |w_n|) > 0. \tag{*}$$

We note that (P_n) is an **increasing sequence of positive numbers**, since,

$\frac{P_{n+1}}{P_n} = 1 + |w_{n+1}| > 1$, meaning that $\boldsymbol{P_{n+1} > P_n}$, $\boldsymbol{\forall n}$, **i.e.** $\boldsymbol{(P_n) \nearrow}$.

From the power series expansion of the exponential function e^x, (see Example 5-2),

$e^x = 1 + \frac{x}{1!} + \frac{x^2}{2!} + \frac{x^3}{3!} + \frac{x^4}{4!} + \cdots$, we obtain, $\boldsymbol{e^x} > 1 + x, \ \forall x > 0.$ (**)

From (*) and (**) we have,

$\boldsymbol{P_n = \prod_{n=1}^{\infty}(1 + |w_n|) < e^{|w_1|+|w_2|+|w_3|+\cdots|w_n|} < e^w}$,

since **by assumption** $\boldsymbol{\sum_{n=1}^{\infty}|w_n|}$ **converges to a finite positive number** $\boldsymbol{w} > 0$.

It turns out that **the sequence** $\boldsymbol{(P_n)}$ **is positive, increasing and bounded above by** $\boldsymbol{e^w}$, therefore the sequence (P_n) is convergent, i.e. $\lim P_n < \infty$, meaning that $\boldsymbol{\prod_{n=1}^{\infty}(1 + w_n) < \infty}$, and this completes the proof.

b) We shall now show that **if** $\boldsymbol{\prod_{n=1}^{\infty}(1 + |w_n|)}$ **converges then** $\boldsymbol{\sum_{n=1}^{\infty}|w_n|}$ **converges.**

Let (S_n) be the sequence of partial sums of the series $\sum_{n=1}^{\infty}|w_n|$, i.e. $S_n = |w_1| + |w_2| + |w_3| + \cdots + |w_n|$.

Obviously the sequence (S_n) is increasing since $\boldsymbol{S_{n+1} > S_n}$, $\boldsymbol{\forall n}$.

At the same time $\boldsymbol{(S_n)}$ **is bounded above**, since

$\boldsymbol{0 < S_n = |w_1| + |w_2| + + \cdots + |w_n| < (1 + |w_1|)(1 + |w_2|) \cdots (1 + |w_n|) < \prod_{n=1}^{\infty}(1 + |w_n|) = P}$.

Therefore **the sequence** $\boldsymbol{(S_n)}$ **being increasing and bounded above is convergent**, i.e. $\boldsymbol{\sum_{n=1}^{\infty}|w_n| < \infty}$.

Theorem 11-3.

If an infinite product is absolutely convergent, then it is convergent. In symbols,

If $\boldsymbol{\prod_{n=1}^{\infty}(1 + |w_n|) < \infty}$**, then** $\boldsymbol{\prod_{n=1}^{\infty}(1 + w_n) < \infty}$.

Proof: Since $\prod_{n=1}^{\infty}(1+|w_n|)<\infty$, the series $\sum_{n=1}^{\infty}|w_n|<\infty$, by Theorem 11-2. We shall now show that **the series $\sum_{n=1}^{\infty}|\ln(1+w_n)|$ is convergent as well.**

Let us consider the two series, $\sum_{n=1}^{\infty}|\ln(1+w_n)|$ and $\sum_{n=1}^{\infty}|w_n|$. (*)

Both series have positive terms, and furthermore,

$$\lim_{n\to\infty}\frac{|\ln(1+w_n)|}{|w_n|}=\left|\lim_{n\to\infty}\frac{\ln(1+w_n)}{w_n}\right|=1.$$

(This is so, since $\lim_{n\to\infty} w_n=0$, (why?), and $\lim_{x\to 0}\frac{\ln(1+x)}{x}=1$).

By virtue of Theorem 5-2, **(the Limit Comparison Test)**, both series in (*) are of the same nature, and **since** $\sum_{n=1}^{\infty}|w_n|<\infty$, the **series** $\sum_{n=1}^{\infty}|\ln(1+w_n)|<\infty$ **as well**, and therefore the series $\sum_{n=1}^{\infty}\ln(1+w_n)<\infty$, by Theorem 7-1, **(every absolutely convergent series is itself convergent)**.

Let $\sum_{n=1}^{\infty}\ln(1+w_n)=S$, where S is a finite number. Then,

$\sum_{n=1}^{\infty}\ln(1+w_n)=S$ implies that $\ln(\prod_{n=1}^{\infty}(1+w_n))=S$, or finally

$\prod_{n=1}^{\infty}(1+w_n)=e^S<\infty$, and this completes the proof.

Note: The converse of Theorem 11-3, **is not necessarily true**, i.e.
if $\prod_{n=1}^{\infty}(1+w_n)<\infty$, this does not necessarily mean that $\prod_{n=1}^{\infty}(1+|w_n|)<\infty$.

For example, let us consider the product

$$\prod_{n=1}^{\infty}\left(1+\frac{(-1)^{n+1}}{n}\right)=\left(1+\frac{1}{1}\right)\left(1-\frac{1}{2}\right)\left(1+\frac{1}{3}\right)\left(1-\frac{1}{4}\right)\cdots=\frac{2}{1}\cdot\frac{1}{2}\cdot\frac{4}{3}\cdot\frac{3}{4}\cdots=1,$$

while $\prod_{n=1}^{\infty}\left(1+\left|\frac{(-1)^{n+1}}{n}\right|\right)=\prod_{n=1}^{\infty}\left(1+\frac{1}{n}\right)=\frac{2}{1}\cdot\frac{3}{2}\cdot\frac{4}{3}\cdots=+\infty$, (see also Problem 11-18).

The following Theorem is also important.

Theorem 11-4.

The infinite product $\prod_{n=1}^{\infty}(1-w_n)$ where $0 \le w_n < 1$, is convergent, if and only if the series $\sum_{n=1}^{\infty} w_n$ is convergent.

Proof: a) Let us assume that $\sum_{n=1}^{\infty} w_n < \infty$. Since $|w_n| = |-w_n| = w_n$, by virtue of Theorem 11-2, the product $\prod_{n=1}^{\infty}(1+|-w_n|) < \infty$, and thus, by Theorem 11-3, $\prod_{n=1}^{\infty}(1-w_n) < \infty$, meaning that the infinite product converges.

b) Let us now assume that $\prod_{n=1}^{\infty}(1-w_n) = P < \infty$. Since by assumption, $0 \le w_n < 1,\ 1+w_n \le \frac{1}{1-w_n}$, therefore,

$$\prod_{n=1}^{\infty}(1+w_n) \le \frac{1}{\prod_{n=1}^{\infty}(1-w_n)} = \frac{1}{P},$$

meaning that the infinite product $\prod_{n=1}^{\infty}(1+w_n) < \infty$, and by Theorem 11-2, the series $\sum_{n=1}^{\infty} w_n$ converges as well.

Example 11-1.

Investigate the convergence of the following infinite products,

a) $\prod_{k=1}^{\infty}\left(1+\frac{1}{\sqrt[3]{k^2}}\right)$ **b)** $\prod_{k=1}^{\infty}\left(1+\frac{1}{\sqrt[2]{k^3}}\right)$ **c)** $\prod_{k=1}^{\infty}\left(1+\frac{2k}{3\sqrt{k^2+1}}\right)$.

Solution

a) In our case, $w_k = \frac{1}{\sqrt[3]{k^2}} = \frac{1}{k^{\left(\frac{2}{3}\right)}} > 0$.

The series $\sum_{k=1}^{\infty}\frac{1}{k^{\left(\frac{2}{3}\right)}}$ is **a p –series with $p = \frac{2}{3}$** < 1, and is therefore divergent, i.e. $\sum_{k=1}^{\infty}\frac{1}{k^{\left(\frac{2}{3}\right)}} = +\infty$, (see Example 5-3). By virtue of Theorem 11-2, the infinite product $\prod_{k=1}^{\infty}\left(1+\frac{1}{\sqrt[3]{k^2}}\right)$ diverges as well.

b) In this case $w_k = \frac{1}{\sqrt{k^3}} = \frac{1}{k^{\left(\frac{3}{2}\right)}} > 0$ and since $\sum_{k=1}^{\infty}\frac{1}{k^{\left(\frac{3}{2}\right)}}$ converges, (is **a p –series with $p = \frac{3}{2}$** > 1), so does the $\prod_{k=1}^{\infty}\left(1+\frac{1}{\sqrt{k^3}}\right)$.

c) The $w_k = \frac{2k}{3\sqrt{k^2+1}}$ and since the $\mathbf{lim}_{k\to\infty} \boldsymbol{w_k} = \frac{2}{3} \neq \mathbf{0}$, the infinite product $\prod_{k=1}^{\infty}\left(1 + \frac{2k}{3\sqrt{k^2+1}}\right)$ is divergent.

Example 11-2.

Investigate for convergence the following infinite products,

a) $\prod_{n=1}^{\infty}(1 + Arccot(2n^2))$ **b)** $\prod_{n=1}^{\infty}\left(1 + \frac{\cos(n\pi)}{n^3+5}\right)$

c) $\prod_{n=1}^{\infty}\left(1 - \frac{1}{\sqrt{n+1}}\right)$.

Solution

a) In this case $w_n = Arccot(2n^2) > 0$ and since $\sum_{n=1}^{\infty} Arccot(2n^2)$ converges to the number $\left(\frac{\pi}{4}\right)$, (see Example 3-6), the infinite product $\prod_{n=1}^{\infty}(1 + Arccot(2n^2)$ is convergent, by Theorem 11-2.

b) The term $w_n = \frac{\cos(n\pi)}{n^3+5} = \frac{(-1)^n}{n^3+5}$ and $|w_n| = \frac{1}{n^3+5}$.

The series $\sum_{n=1}^{\infty}|w_n| = \sum_{n=1}^{\infty}\frac{1}{n^3+5}$ is convergent (why?), therefore $\prod_{n=1}^{\infty}(1 + |w_n|)$ is convergent, (Theorem 11-2), which in turn implies that $\prod_{n=1}^{\infty}(1 + w_n) = \prod_{n=1}^{\infty}\left(1 + \frac{\cos(n\pi)}{n^3+5}\right)$ is convergent as well, by virtue of Theorem 11-3.

c) The term $w_n = \frac{1}{\sqrt{n+1}}$ and obviously $0 < \frac{1}{\sqrt{n+1}} < 1$.

The series $\sum_{n=1}^{\infty}\frac{1}{\sqrt{n+1}} = +\infty$ (why?) and application of Theorem 11-4 yields that $\prod_{n=1}^{\infty}(1 - w_n) = \prod_{n=1}^{\infty}\left(1 - \frac{1}{\sqrt{n+1}}\right)$ is divergent.

Example 11-3.

Evaluate the infinite product $\prod_{k=1}^{\infty}\left(1 + \frac{1}{k(k+2)}\right)$.

Solution

Since $\sum_{k=1}^{\infty}\frac{1}{k(k+2)} < \infty$ (why?), the infinite product **is convergent**.

If we call $u_k = 1 + \frac{1}{k(k+2)} = \frac{k^2+2k+1}{k(k+2)} = \frac{(k+1)^2}{k(k+2)} = \frac{k+1}{k} \cdot \frac{k+1}{k+2}$,

the $n^{\underline{th}}$ partial product P_n will be,

$$P_n = \prod_{k=1}^{n} u_k = \prod_{k=1}^{n} \left(\frac{k+1}{k} \cdot \frac{k+1}{k+2}\right) = \left(\prod_{k=1}^{n} \frac{k+1}{k}\right)\left(\prod_{k=1}^{n} \frac{k+1}{k+2}\right), \text{ or}$$

$$P_n = \left(\frac{2}{1} \cdot \frac{3}{2} \cdot \frac{4}{3} \cdots \frac{n}{n-1} \cdot \frac{n+1}{n}\right)\left(\frac{2}{3} \cdot \frac{3}{4} \cdot \frac{4}{5} \cdots \frac{n}{n+1} \cdot \frac{n+1}{n+2}\right) = (n+1)\frac{2}{n+2} = 2\frac{n+1}{n+2},$$

and finally,

$$\prod_{k=1}^{\infty} \left(1 + \frac{1}{k(k+2)}\right) = \lim_{n\to\infty} P_n = \lim_{n\to\infty} \left(2\frac{n+1}{n+2}\right) = 2 \cdot 1 = 2.$$

Example 11-4.

Show that $\prod_{k=1}^{\infty} \left(1 - \frac{x^2}{k^2}\right)$ converges for any $x \in \mathbb{R}$.

Solution

If we set $w_k = -\frac{x^2}{k^2}$, then $|w_k| = \frac{x^2}{k^2}$ and the series $\sum_{k=1}^{\infty} |w_k| = x^2 \sum_{k=1}^{\infty} \frac{1}{k^2}$ converges **for any fixed value of x**, since the series $\sum_{k=1}^{\infty} \frac{1}{k^2}$ is **a p −series, with $p = 2 > 1$.**

By Theorem 11-2 the infinite product is absolutely convergent, and therefore by Theorem 11-3 the product $\prod_{k=1}^{\infty} \left(1 - \frac{x^2}{k^2}\right)$ converges for any real value of x.

Note: If x is replaced by $\frac{x}{\pi}$, we obtain the infinite product $\prod_{k=1}^{\infty} \left(1 - \frac{x^2}{k^2\pi^2}\right)$, which obviously converges **for all real values of x**, $(-\infty < x < \infty)$. **(**As a matter of fact, it can be shown that **this infinite product is convergent if x is replaced by any arbitrary complex number z**).

This means that for each $x \in \mathbb{R}$, we have a value for the infinite product $\prod_{k=1}^{\infty} \left(1 - \frac{x^2}{k^2\pi^2}\right)$, and **this correspondence defines a function $f(x)$**, i.e. $f(x) = \prod_{k=1}^{\infty} \left(1 - \frac{x^2}{k^2\pi^2}\right)$.

It can be shown that $f(x) = \frac{\sin x}{x}$, i.e.

$$\prod_{k=1}^{\infty}\left(1-\frac{x^2}{k^2\pi^2}\right)=\frac{\sin x}{x}.$$

This formula, again, is due to **Euler**, who used it in order to find the sum of the series $\sum_{k=1}^{\infty}\frac{1}{k^2}=\frac{1}{1^2}+\frac{1}{2^2}+\frac{1}{3^2}+\frac{1}{4^2}+\cdots$.

This is the famous **"Basel's Problem"**, solved for the first time by the great Mathematician **Leonard Euler, in 1730**, (see also Example 5-3). Euler's method characterized by great ingenuity, opened a whole new era in analysis. We shall describe, below, in loose terms, Euler's method for the evaluation of $\sum_{k=1}^{\infty}\frac{1}{k^2}$.

Example 11-5. (The Basel's Problem)

Find the sum of the series $\sum_{k=1}^{\infty}\frac{1}{k^2}=\frac{1}{1^2}+\frac{1}{2^2}+\frac{1}{3^2}+\frac{1}{4^2}+\cdots$.

Solution

i) The series $\sum_{k=1}^{\infty}\frac{1}{k^2}$ is a p –series with $p=2>1$, therefore converges to a finite positive number S which is to be found.

Starting with the series expansion for $\sin x$, given in (8-25), one easily obtains an expression for the function $g(x)=\frac{\sin x}{x}$, i.e.

$$g(x)=\frac{\sin x}{x}=1-\frac{1}{3!}x^2+\frac{1}{5!}x^4-\frac{1}{7!}x^6+\frac{1}{9!}x^8-\frac{1}{11!}x^{10}+\cdots,\quad x\in\mathbb{R}. \qquad (*)$$

(Note that $g(0)=\lim_{x\to 0}\frac{\sin x}{x}=1$).

Euler's **original idea** was to consider the right side in (*), as **"a polynomial in x, of infinite degree".** The roots of this polynomial, will actually be **the roots of $\sin x=0$**, except the root $x=0$, since $g(0)=1$. In other words the roots of $g(x)$ will be,

$$\{\cdots,-3\pi,\ -2\pi,\ -\pi,\ \pi,\ 2\pi,\ 3\pi,\cdots\}.$$

ii) Euler knew that any polynomial of the form,

$$f(x)=1+a_1x+a_2x^2+\cdots+a_nx^n \quad \text{having roots } r_1,r_2,r_3,\cdots,r_n \qquad (**)$$

can be expressed, in terms of the roots, as

$$\boldsymbol{f(x) = \left(1-\frac{x}{r_1}\right)\left(1-\frac{x}{r_2}\right)\left(1-\frac{x}{r_3}\right)\cdots\left(1-\frac{x}{r_n}\right) = \prod_{k=1}^{n}\left(1-\frac{x}{r_k}\right)} \qquad (***)$$

Indeed, both expressions (**) and (***), represent **two $n^{\underline{th}}$ degree polynomials, coinciding at $(n+1)$ points, i.e. at the points** $\boldsymbol{x = 0, r_1, r_2, r_3, \cdots, r_n}$. Therefore, (**) and (***) **are identical expressions.**

iii) Euler did not hesitate to apply this simple idea, to the **"infinite degree polynomial"** in (*), to obtain,

$$g(x) = \frac{\sin x}{x} = \cdots\left(1+\frac{x}{3\pi}\right)\left(1+\frac{x}{2\pi}\right)\left(1+\frac{x}{\pi}\right)\left(1-\frac{x}{\pi}\right)\left(1-\frac{x}{2\pi}\right)\left(1-\frac{x}{3\pi}\right)\cdots,$$

or even more,

$$\boldsymbol{g(x) = \frac{\sin x}{x} = \left(1-\frac{x^2}{\pi^2}\right)\left(1-\frac{x^2}{(2\pi)^2}\right)\left(1-\frac{x^2}{(3\pi)^2}\right)\cdots =}$$

$$\boldsymbol{= \prod_{n=1}^{\infty}\left(1-\frac{x^2}{(n\pi)^2}\right)} \qquad (****)$$

iv) The coefficient of x^2 in this infinite product is

$$-\left(\frac{1}{\pi^2}+\frac{1}{(2\pi)^2}+\frac{1}{(3\pi)^2}+\cdots\right) = -\frac{1}{\pi^2}\left(\frac{1}{1^2}+\frac{1}{2^2}+\frac{1}{3^2}+\cdots\right) = -\frac{1}{\pi^2}\sum_{n=1}^{\infty}\frac{1}{n^2}.$$

On the other hand, the coefficient of x^2, in (*) is $\left(-\frac{1}{3!}\right) = -\frac{1}{6}$.

Equating the coefficients of x^2, in both sides, we get,

$$\boldsymbol{-\frac{1}{\pi^2}\sum_{n=1}^{\infty}\frac{1}{n^2} = -\frac{1}{6} \quad or \quad \sum_{n=1}^{\infty}\frac{1}{n^2} = \frac{\pi^2}{6}}.$$

A beautiful, interesting and quite unexpected result.

Note: Using similar arguments, one may show that

$$\boldsymbol{\sum_{n=1}^{\infty}\frac{1}{n^4} = \frac{\pi^4}{90} \quad and \quad \sum_{n=1}^{\infty}\frac{1}{n^6} = \frac{\pi^6}{945}}.$$

It is of interest to note that the **sum of odd powers of the reciprocals of all the positive integers**, i.e. the sum of series of the form

$$\sum_{n=1}^{\infty}\frac{1}{n^3}, \quad \sum_{n=1}^{\infty}\frac{1}{n^5}, \quad \sum_{n=1}^{\infty}\frac{1}{n^7}, etc.$$

is not known **in closed form**. It is still **an open problem**.

PROBLEMS

11-1) Investigate for convergence the following infinite products,

a) $\prod_{k=2}^{\infty}\left(1-\frac{1}{k^5}\right)$ **b)** $\prod_{k=1}^{\infty}\left(1+\frac{1}{\sqrt[7]{k^8}}\right)$ **c)** $\prod_{k=1}^{\infty}\left(1+\frac{2k^3}{7k^3+5}\right)$.

(**Answer: a)** Converges, **b)** Converges, **c)** Divergence).

11-2) Show that $\prod_{k=2}^{\infty}\frac{k^3-1}{k^3+1}=\frac{2}{3}$.

Hint: The general term $\frac{k^3-1}{k^3+1}=\frac{(k-1)(k^2+k+1)}{(k+1)(k^2-k+1)}=\frac{(k-1)(k-r_1)(k-r_2)}{(k+1)(k-x_1)(k-x_2)}$ where r_1 and r_2 are the roots of $k^2+k+1=0$, while x_1 and x_2 are the roots of $k^2-k+1=0$. Note that $x_1=r_1+1,\ x_2=r_2+1$, etc.

11-3) Show that $\prod_{n=5}^{\infty}\left(1-\frac{2}{n(n+1)}\right)=\frac{2}{3}$.

Hint: $u_n=1-\frac{2}{n(n+1)}=\frac{n^2+n-2}{n(n+1)}=\frac{(n-1)(n+2)}{n(n+1)}=\frac{n-1}{n}\cdot\frac{n+2}{n+1}$.

11-4) Consider a convergent series $\sum_{k=1}^{\infty}\boldsymbol{u_k}=\boldsymbol{S}$, and let (s_n) be the sequence of partial sums corresponding to this series. Show that the infinite product $\prod_{k=2}^{\infty}\left(1+\frac{u_k}{s_{k-1}}\right)$ converges to $\frac{\mathbf{1}}{\boldsymbol{u_1}}\sum_{k=1}^{\infty}\boldsymbol{u_k}=\frac{\boldsymbol{S}}{\boldsymbol{u_1}}$

11-5) Making use of Problem 11-4 show that $\prod_{k=2}^{\infty}\left(1+\frac{1}{2^k-2}\right)=2$.

Hint: Consider $u_k=\frac{1}{2^k}$.

11-6) Show that $\prod_{n=1}^{k}\left(1-\left(\tan\left(\frac{x}{2^n}\right)\right)^2\right)=2^k\frac{\left(\tan\left(\frac{x}{2^k}\right)\right)^2}{\tan x}$.

Hint: $1-(\tan x)^2=\frac{2\tan x}{\tan 2x}$.

11-7) Show that $\prod_{n=1}^{\infty}\left(1-\left(\tan\left(\frac{x}{2^n}\right)\right)^2\right)=\frac{x}{\tan x}$.

Hint: The $\lim_{y\to 0}\frac{\tan y}{y}=\lim_{y\to 0}\left\{\frac{\sin y}{y}\cdot\frac{1}{\cos y}\right\}=1\cdot 1=1$.

11-8) Show that $\prod_{n=1}^{k} \frac{1+\left(\cot\left(\frac{x}{2^n}\right)\right)^2}{-1+\left(\cot\left(\frac{x}{2^n}\right)\right)^2} = 2^k \frac{\sin\left(\frac{x}{2^{k-1}}\right)}{\sin 2x}$.

Hint: $\frac{1+(\cot x)^2}{-1+(\cot x)^2} = \frac{1}{\cos 2x} = \frac{2\sin 2x}{\sin 4x}$

11-9) Find the infinite product $\prod_{n=1}^{\infty} \frac{1+\left(\cot\left(\frac{x}{2^n}\right)\right)^2}{-1+\left(\cot\left(\frac{x}{2^n}\right)\right)^2}$.

(**Answer:** $\frac{2x}{\sin 2x}$).

11-10) Making use of the **Euler's Product formula** in Example 11-4, show **the Wallis formula,**

$$\frac{\pi}{2} = \frac{2\cdot 2\cdot 4\cdot 4\cdot 6\cdot 6\cdot 8\cdot 8\cdots}{1\cdot 3\cdot 3\cdot 5\cdot 5\cdot 7\cdot 7\cdot 9\cdots} = \prod_{k=1}^{\infty} \frac{(2k)^2}{(2k-1)(2k+1)}.$$

11-11) Investigate for convergence the following infinite products,

a) $\prod_{k=3}^{\infty}\left(1-\frac{1}{\sqrt{k+5}}\right)$, **b)** $\prod_{k=1}^{\infty}\left(1+\frac{\sqrt[3]{k^4}}{\sqrt[3]{2k^4+7k^2+1}}\right)$,

c) $\prod_{k=10}^{\infty}\left(1-\frac{(-1)^k}{k^5}\right)$, **d)** $\prod_{k=1}^{\infty}\left(1+\frac{x}{k^3+7}\right)$, where $x \in \mathbb{R}$.

(**Answer: a)** Diverges, **b)** Diverges, **c)** Converges and **d)** Converges for all real x).

11-12) a) If n is a positive integer, show that

$$\cos\frac{\pi}{2^{n+1}} = \frac{1}{2}\cdot\sqrt{2+\sqrt{2+\sqrt{2+\sqrt{2+\cdots+\sqrt{2}}}}}$$

where in the right side there are n **successive radicals.**

b) Making use of part (a), show that

$$\lim_{n\to\infty}\left\{\frac{1}{2^n}\cdot\prod_{k=1}^{n}\left(\sqrt{2+\sqrt{2+\sqrt{2+\cdots+\sqrt{2}}}}\right)\right\} = \frac{2}{\pi}$$

where inside the parentheses there are k **successive radicals.**

Hint: Make use of the Trigonometric identity, $\cos x = \frac{\sin 2x}{2 \sin x}$ and the fact that $\lim_{y \to 0} \frac{\sin y}{y} = 1$.

11-13) If $x \in \mathbb{R}$, show that $\boldsymbol{\cos x = \prod_{n=1}^{\infty} \left(1 - \frac{4x^2}{(2n-1)^2 \pi^2}\right)}$.

Hint: Make use of the identity $\cos x = \frac{\sin 2x}{2 \sin x} = \frac{\sin 2x \div 2x}{\sin x \div x}$ and then express the $\frac{\sin 2x}{2x}$ and the $\frac{\sin x}{x}$ according to **the Euler's product formula**, (see Example 11-4).

11-14) Making use of **the Euler's product formula** for the function $\frac{\sin x}{x}$ and the equations (8-34) and (8-35), show that for any $x \in \mathbb{R}$,

$$\boldsymbol{\sinh x = x \prod_{n=1}^{\infty} \left(1 + \frac{x^2}{n^2 \pi^2}\right)} \quad \text{and} \quad \boldsymbol{\cosh x = \prod_{n=1}^{\infty} \left(1 + \frac{4x^2}{(2n-1)^2 \pi^2}\right)}.$$

Note: The same formulas apply, if the real x is replaced by **any complex number** z.

11-15) a) Show that $\prod_{n=1}^{\infty} \left(1 + \frac{1}{n^k}\right)$ is convergent for any $k > 1$.

b) Show that $\prod_{n=2}^{\infty} \left(1 - \frac{1}{n^k}\right)$ is absolutely convergent and therefore convergent, for any $k > 1$.

c) For $k = 1$, show that the infinite product in (b), **diverges to zero**.

11-16) If $|x| < 1$, show that the infinite product $\prod_{n=1}^{\infty} (1 + x^n)$ is absolutely convergent.

Hint: Consider the series $\sum |x^n| = \sum |x|^n$, and make use of Theorem 11-2.

11-17) Consider the infinite product $\prod_{n=1}^{\infty} \left(1 + x^{2^n}\right)$ where $|x| < 1$. Show that the product is convergent and find its value.

(Answer: $\frac{1}{1-x^2}$).

11-18) Show that the product $\prod_{n=1}^{\infty} \left(1 + \frac{1}{n}\right)$ diverges to $+\infty$, while the product $\prod_{n=2}^{\infty} \left(1 - \frac{1}{n}\right)$ diverges to zero.

11-19) Find the values of $x \in \mathbb{R}$, for which the product $\prod_{n=1}^{\infty}\left(1+\frac{(x+2)^n}{n}\right)$ is convergent.

Hint: It suffices to find the values of x, for which the series $\sum_{n=1}^{\infty}\frac{(x+2)^n}{n}$ is absolutely convergent. At the end points of the convergence interval, additional investigation is needed.

(**Answer:** $-3 \leq x < -1$).

11-20) If the series $\sum u_n$ is absolutely convergent, show that the infinite product $\prod(1+xu_n)$ will be absolutely convergent as well, for any value of $x \in \mathbb{R}$.

11-21) If $x \in \mathbb{R}-\{0\}$ evaluate the product $\prod_{k=1}^{n}\cosh\frac{x}{2^n}$.

Hint: With the aid of (8-18) and (8-19), show that $\mathbf{\sinh 2t = 2\sinh t\cosh t}$, or $\cosh t = \frac{\sinh 2t}{2\sinh t}$. Setting $t=\frac{x}{2^n}$, we have $\mathbf{\cosh\left(\frac{x}{2^n}\right) = \frac{\sinh\left(\frac{x}{2^{n-1}}\right)}{2\sinh\left(\frac{x}{2^n}\right)}}$, etc.

(**Answer:** $\frac{\sinh x}{x}\cdot\frac{\left(\frac{x}{2^n}\right)}{\sinh\left(\frac{x}{2^n}\right)}$).

11-22) a) Show that $\lim_{y\to 0}\frac{\sinh y}{y}=1$.

b) Show that $\prod_{n=1}^{\infty}\cosh\left(\frac{x}{2^n}\right)=\frac{\sinh x}{x}$.

Hint: a) By definition, $\mathbf{\sinh y = \frac{e^y - e^{-y}}{2}}$. Express e^y and e^{-y} by their series expansions, as given in (8-16) and (8-17) respectively, form the ratio $\frac{\sinh y}{y}$ and then pass to the limit as $y \to 0$.

b) Use the result obtained in Problem 11-21, note that as $n \to \infty$, the quantity $\left(\frac{x}{2^n}\right) \to \mathbf{0}$ for **an arbitrary but fixed** $\boldsymbol{x}$,and then make use of part (a).

11-23) Starting with the result obtained in Problem 11-22 (b), and setting $\boldsymbol{x} = \ln \boldsymbol{t}$, (where $t \in \mathbb{R}^+ - \{1\}$), show that,

$$\frac{\ln t}{t-1} = \prod_{n=1}^{\infty} \frac{2}{1+\sqrt[2^n]{t}} = \frac{2}{1+\sqrt{t}} \cdot \frac{2}{1+\sqrt[4]{t}} \cdot \frac{2}{1+\sqrt[8]{t}} \cdots.$$

11-24) Prove the following infinite product expression for $\ln 2$,

$$\ln 2 = \frac{2}{1+\sqrt{2}} \cdot \frac{2}{1+\sqrt[4]{2}} \cdot \frac{2}{1+\sqrt[8]{2}} \cdots.$$

Hint: Apply the general formula obtained in the previous problem at $t = 2$.

11-25) Making use of Problem 11-14, show that

$$\prod_{n=1}^{\infty} \left(1 + \frac{1}{n^2}\right) = \frac{\sinh \pi}{\pi} = \frac{e^{\pi} - e^{-\pi}}{2\pi}.$$

Note: If the $+\frac{1}{n^2}$ term is replaced by $-\frac{1}{n^2}$, the value of the infinite product $\prod_{n=2}^{\infty}\left(1 - \frac{1}{n^2}\right) = \frac{1}{2}$.

11-26) Show that

$$\pi = \frac{\prod_{n=1}^{\infty}\left(1 + \frac{1}{4n^2 - 1}\right)}{\sum_{n=1}^{\infty} \frac{1}{4n^2 - 1}}.$$

Hint: Making use of **the Wallis formula**, Problem 11-10 the $\frac{\pi}{2} = \prod_{n=1}^{\infty} \frac{4n^2}{4n^2-1} = \prod_{n=1}^{\infty}\left(1 + \frac{1}{4n^2-1}\right)$, while the series $\sum_{n=1}^{\infty} \frac{1}{4n^2-1}$ is a telescoping series converging to $\frac{1}{2}$, (show it).

11-27) Starting with the **Wallis formula**, show that $\sqrt{\pi} = \lim_{n\to\infty} \frac{(n!)^2 2^{2n}}{(2n)!\sqrt{n}}$.

11-28) a) If $x \neq 0$, show that $\sum_{k=1}^{\infty} \frac{1}{(x+k-1)(x+k)} = \frac{1}{x}$.

b) With the aid of Problem 11-4, show that $\prod_{k=2}^{\infty}\left(1 + \frac{x}{(k+x)(k-1)}\right) = x + 1$.

Hint: The series in part (a) is a telescoping series.

11-29) In Example 5-3, we introduced **the Riemann's zeta function,**

$$\zeta(x) = \sum_{n=1}^{\infty} \frac{1}{n^x}, \quad x > 1.$$

Euler again, gave another infinite product representation for $\zeta(x)$, in terms of **all the positive prime numbers**.

Show the following **infinite product representation for $\zeta(x)$**,

$$\zeta(x) = \sum_{n=1}^{\infty} \frac{1}{n^x} = \frac{1}{1^x} + \frac{1}{2^x} + \frac{1}{3^x} + \cdots = \frac{1}{\prod_{p=prime}\left(1 - \frac{1}{p^x}\right)}, \qquad x > 1,$$

where the **infinite product extends over all the prime numbers $p \in \mathbb{N}$**, i.e.

$$\prod_{p=prime} \left(1 - \frac{1}{p^x}\right) = \left(1 - \frac{1}{2^x}\right)\left(1 - \frac{1}{3^x}\right)\left(1 - \frac{1}{5^x}\right)\left(1 - \frac{1}{7^x}\right)\left(1 - \frac{1}{11^x}\right)\cdots.$$

Hint: $\zeta(x) = \frac{1}{1^x} + \frac{1}{2^x} + \frac{1}{3^x} + \frac{1}{4^x} + \frac{1}{5^x} + \frac{1}{6^x} + \frac{1}{7^x} + \frac{1}{8^x} + \frac{1}{9^x} + \frac{1}{10^x} + \cdots$, or

$$\frac{1}{2^x}\zeta(x) = \frac{1}{2^x} + \frac{1}{4^x} + \frac{1}{6^x} + \frac{1}{8^x} + \frac{1}{10^x} + \frac{1}{12^x} + \frac{1}{14^x} + \frac{1}{16^x} + \frac{1}{18^x} + \frac{1}{20^x} + \cdots,$$

and subtracting the second from the first we get,

$$\left(1 - \frac{1}{2^x}\right)\zeta(x) = \frac{1}{1^x} + \frac{1}{3^x} + \frac{1}{5^x} + \frac{1}{7^x} + \frac{1}{9^x} + \frac{1}{11^x} + \frac{1}{13^x} + \frac{1}{15^x} + \frac{1}{17^x} + \frac{1}{19^x} + \cdots.$$

At this step, **we have removed all the multiples of 2**.

Next we multiply both sides of this equation by $\frac{1}{3^x}$, to obtain,

$$\frac{1}{3^x}\left(1 - \frac{1}{2^x}\right)\zeta(x) = \frac{1}{3^x} + \frac{1}{9^x} + \frac{1}{15^x} + \frac{1}{21^x} + \frac{1}{27^x} + \frac{1}{33^x} + \frac{1}{39^x} + \frac{1}{45^x} + \frac{1}{51^x} + \frac{1}{57^x} + \cdots,$$

and subtracting again the last two equations term wise, we get

$$\left(1 - \frac{1}{2^x}\right)\left(1 - \frac{1}{3^x}\right)\zeta(x) = \frac{1}{1^x} + \frac{1}{5^x} + \frac{1}{7^x} + \frac{1}{11^x} + \frac{1}{13^x} + \frac{1}{17^x} + \frac{1}{19^x} + \cdots.$$

At this step, **we have removed all the multiples of 3**.

Proceeding similarly, we obtain (**after an infinite number of steps**),

$$\left(1 - \frac{1}{2^x}\right)\left(1 - \frac{1}{3^x}\right)\left(1 - \frac{1}{5^x}\right)\left(1 - \frac{1}{7^x}\right)\left(1 - \frac{1}{11^x}\right)\cdots\zeta(x) = 1,$$

from which the desired result follows immediately.

11-30) Show that $\prod_{p=prime}\left(1-\frac{1}{p^2}\right)=\frac{6}{\pi^2}$.

Hint: $\prod_{p=prime}\left(1-\frac{1}{p^2}\right)=\frac{1}{\zeta(2)}=\frac{1}{\left(\frac{\pi^2}{6}\right)}=\frac{6}{\pi^2}$,since $\zeta(2)=\sum_{n=1}^{\infty}\frac{1}{n^2}=\frac{\pi^2}{6}$

(see Example 11-5).

Made in the USA
Las Vegas, NV
09 May 2021

22729503R00083